BIOCHEMICAL SOCIETY SYMPOSIA

No. 41

BIOCHEMICAL ADAPTATION
TO
ENVIRONMENTAL CHANGE

BIOCHEMICAL SOCIETY SYMPOSIUM No. 41
held at the University of Liverpool, July 1975

Biochemical Adaptation
to
Environmental Change

ORGANIZED BY

R. M. S. SMELLIE

AND

EDITED BY

R. M. S. SMELLIE AND J. F. PENNOCK

1976
LONDON: THE BIOCHEMICAL SOCIETY

The Biochemical Society,
7 Warwick Court,
London WC1R 5DP, U.K.

ISBN 0 904498 01 8
ISSN 0067-8694

*Printed in Great Britain by William Clowes & Sons Limited,
London, Colchester and Beccles*

List of Contributors

D. E. Atkinson (*Biochemistry Division, Department of Chemistry, University of California at Los Angeles, Los Angeles, CA 90024, U.S.A.*)

A. S. Blix (*Institute of Medical Biology, Section of Physiology, University of Tromsø, Tromsø, Norway*)

B. Crabtree (*Department of Animal Physiology and Nutrition, University of Leeds, Vicarage Terrace, Leeds LS2 9JT, U.K.*)

A. de Zwaan (*Laboratory of Chemical Animal Physiology, State University of Utrecht Transitorium III, 8 Padualaan, Utrecht, The Netherlands*)

C. J. Duncan (*Department of Zoology, University of Liverpool, P.O. Box 147, Liverpool L69 3BX, U.K.*)

P. W. Hochachka (*Department of Zoology, University of British Columbia, Vancouver 8, B.C., Canada*)

J. H. F. M. Kluytmans (*Laboratory of Chemical Animal Physiology, State University of Utrecht Transitorium III, 8 Padualaan, Utrecht, The Netherlands*)

P. S. Low (*Scripps Institution of Oceanography, University of California, P.O. Box 1529, La Jolla, CA 92093, U.S.A.*)

E. A. Newsholme (*Department of Biochemistry, University of Oxford, South Parks Road, Oxford OX1 3QU, U.K.*)

B. Sacktor (*Gerontology Research Center, National Institute of Aging, National Institutes of Health, Baltimore City Hospitals, Baltimore, MD 21224, U.S.A.*)

E. Schoffeniels (*Laboratoire de Biochimie Générale et Comparée, Université de Liège, 17 Place Delcour, B-4020 Liège, Belgium*)

M. W. Smith (*Agricultural Research Council Institute of Animal Physiology, Babraham, Cambridge CB2 4AT, U.K.*)

G. N. Somero (*Scripps Institution of Oceanography, University of California, P.O. Box 1529, La Jolla, CA 92093, U.S.A.*)

D. I. Zandee (*Laboratory of Chemical Animal Physiology, State University of Utrecht Transitorium III, 8 Padualaan, Utrecht, The Netherlands*)

Preface

This Symposium was conceived largely by Professor Goodwin and Dr. Hochachka with the object of bringing together speakers with expert knowledge of different aspects of the biochemical adaptation to environmental change.

Like many sciences, biochemistry has moved through several phases—firstly a broad descriptive period in which the main features of the structures and metabolic pathways in living systems were studied. This was followed by a second phase in which details of biochemical structure and metabolism were established. Thereafter biochemists became increasingly concerned with mechanisms of control at the subcellular and cellular level, and they are now returning to considerations of biochemical control and the integration of biochemical pathways at the level of whole animals.

Topics in this Symposium range from consideration of the effects of temperature and pressure on enzyme evolution and function to the modifications in the classical pathways of metabolism that have been developed to meet the requirements of particular environments. The species considered include squid, molluscs, birds, seals and insects. Perhaps more than any of its predecessors, this Symposium serves to highlight the advantages of interdisciplinary approaches to the solution of biological problems.

The Biochemical Society is grateful to the speakers who have contributed to the Symposium, and particularly to Dr. Hochachka, who acted not only as a speaker, but also as a Chairman and the source of many of the ideas about the programme.

R. M. S. SMELLIE

Institute of Biochemistry,
University of Glasgow,
Glasgow G12 8QQ,
U.K.

Contents

Contents

Biochem. Soc. Symp. (1976) **41**, 1–2
Printed in Great Britain

Chairman's Introduction

By C. J. DUNCAN

Department of Zoology, University of Liverpool, P.O. Box 147,
Liverpool L69 3BX, U.K.

The concept that animals and plants are morphologically and behaviourally adapted to their environment was fundamental to the thinking of Darwin and Huxley. Inevitably, the ideas of physiological and biochemical adaptation came later and I am sure that many people here today, like me, were strongly influenced as students by Baldwin's little book [*An Introduction to Comparative Biochemistry* (1937) Cambridge University Press, Cambridge]. It opened my eyes to the fascinating world of comparative biochemistry, the chemistry of animal pigments and the classical studies of Needham on nitrogen excretion. Such biochemical adaptations represent responses to changes in the environment on an evolutionary time-scale and several of the papers in this symposium follow in this classic vein. At a time when cell physiology is emphasizing the underlying identities in biochemical organization throughout the animal and plant kingdoms, it is important to remember the significance of biochemical specializations for different animal life-styles. Furthermore, in human terms, the time-scale of evolutionary adaptation need not be long; a rapid rate of reproduction coupled with a short generation time ensures that resistance to antibiotics and pesticides can quickly develop—a new and important field of biochemical and genetical research.

The other form of adaptation shown by living systems is that which occurs during the life-time of an individual. Such changes are termed acclimations (when they are in response to a single, controlled environmental variable in the laboratory) or acclimatizations (when the response occurs under natural environmental conditions), to distinguish them from the evolutionary adaptations described above. It is perhaps important to emphasize these distinctions at the beginning of our symposium. My own interest lies in thermal acclimation, and although we remember, for example, the distinguished early work of Precht, there is no doubt that studies in the whole field of acclimation have mushroomed in recent years. As well as being a fascinating area of study in itself, as many papers today and tomorrow will show, I believe that acclimation represents a valuable experimental tool that is available to biochemists. For many of us, the aim is to translate biochemical results into the context of the physiology of the intact cell. Problems in membrane physiology are an example and one experimental approach involves perturbation of the system by drugs or physicochemical techniques and a subsequent study of the changes in the biochemistry or physiology of the cell. Such methods inevitably have shortcomings, and parallel studies, employing more naturally induced changes, are invaluable. Examples are legion and one remembers studies using material which exhibits genetic modifications, such as the mutants of *Escherichia coli* and the erythrocytes of patients suffering from hereditary sphaerocytosis. Another valuable method is the study of the response to acclimation to a carefully controlled environmental parameter, as for example the dietary manipulation of the membrane lipids of micro-organisms. I am certain that many people, both those interested in the control mechanisms of biochemical processes and those studying the biochemistry of cellular processes, will find new experimental approaches to their problems in the papers in this symposium.

Biochem. Soc. Symp. (1976) **41**, 3–31
Printed in Great Britain

Design of Metabolic and Enzymic Machinery to Fit Lifestyle and Environment

By P. W. HOCHACHKA

*Department of Zoology, University of British Columbia, Vancouver 8, B.C.,
Canada*

Synopsis

Biochemical theory, spectacular as its growth has been in the last half-century,
finds its experimental basis in a remarkably few organisms. The rat, the ox, the
pig, the hen plus a few micro-organisms constitute the primary source of biochemistry's experimental material, while, by contrast, modern systematics
estimates a bewildering and often overlooked diversity of over 10^6 animal species.
As a result of this highly limited 'sampling' of nature, classical biochemistry has
supplied an equally limited insight into how far biochemical mechanisms can be
extended or adapted for (1) allowing an organism's exploitation of its environment
and (2) accommodating environmental change. The situation has improved in
more recent years, so that it is now possible to decipher at least the broad strategic
outlines of biochemical adaptation to the environment. This essay focuses particular attention on the influence of O_2 availability on muscle metabolic organization and control. By drawing examples of muscle metabolism spanning a range
from obligate-anaerobic to obligate-aerobic organization, it is possible to
demonstrate that certain kinds of metabolic features are highly conserved,
whereas others are changed time and time again, in different combinations and
permutations. The latter properties constitute the 'raw material' for adapting
metabolism to specific O_2 regimes, and include (1) the kinds and amounts of
enzymes present in muscle, (2) the fine regulatory circuitry allowing for controlled
transition from low to high work loads and (3) compensation mechanisms for the
adjustment of enzyme reactions with respect to such external parameters as
temperature, pressure, salinity and so forth. Within the inescapable limitations
of phylogenetic origins and the time available for responding to specific environmental conditions, adequate adaptations of the fundamental design of muscle
metabolism have evolved to exploit environments differing greatly in O_2 availability and to accommodate a wide range of muscle functions.

Introduction

One of the difficulties in understanding biochemical adaptation arises from an
odd characteristic: when it is most effective it tends to become most invisible
and difficult to recognize. I like to illustrate this difficulty by considering related
organisms, teleosts or squids, for example, living in different positions in the
marine water column. Despite large differences in the external environment
impinging on their cell chemistry, these organisms may be remarkably similar in

their macroscopic attributes, their hydrodynamics, swimming, growth, respiration, ion and osmoregulation and so forth. Essential adaptations are neither seen nor suspected, and one could well imagine that as ectotherms they may be disobeying certain laws of physics and chemistry. Such is not the case of course. When one examines and compares such organisms at the level of cell chemistry and cell metabolism one appreciates that the visible macroscopic similarities are achieved by fundamental biochemical microscopic differences and specializations. Because these specializations are the outcome of cycles of mutation and selection (i.e. of biochemical adaptation), they are in a proper sense designed systems. Their origin and nature cannot be predicted from basic chemical and physical principles any more readily than can be the shape of fast swimming tunas from knowledge, however complete, of the physics and chemistry of proteins, carbohydrates and fats. What is required for understanding and prediction of designed systems (whether designed by man or by adaptational processes) is knowledge of the rules or principles of design in each case. Armed with these, a shipbuilder can predict (with an accuracy limited only by the completeness of his design rules) the speed, stress and strain consequences of change in hull shape, in polymer coating or in sail position; similarly, a biomechanic can predict the consequences to speed and cost of swimming of change in shape of a tuna's tail (Alexander, 1967). A biologist may have more difficulty unravelling the rules of design, but once this barrier is overcome, his predictive abilities may be better than an engineer's because in living systems, function and design evolve together. Overdesign and underdesign are both strongly selected against, unlike the situation in Detroit, where engineers overdesigned their cars in terms of appearance but underdesigned them in terms of performance [see Atkinson (1976) for a further discussion of this point]. Not surprisingly, biology frequently refers to 'optimization' or to 'optimal design' of organisms and their parts, 'optimal' being defined as the best of possible compromises made by an organism facing conflicting variables. Similar optimization appears to occur at the biochemical as well as the structural levels of organization, and can be well illustrated by recent studies of the functional design of the binding sites of acetylcholinesterase (EC 3.1.1.7).

Acetylcholinesterase, catalysing the hydrolysis of acetylcholine to choline plus acetate, is ubiquitously distributed in the nervous systems of animals. The enzyme, which has been extensively studied, displays a substrate-binding site consisting of two domains: an anionic and an esteratic site. The first is primarily concerned with binding whereas the second is the true catalytic site, the site at which hydrolysis occurs. Both coulombic and hydrophobic interactions contribute to binding (Froede & Wilson, 1972). Since both binding contributions derive from noncovalent 'weak' chemical bonds, which are well known to be highly if differentially sensitive to temperature and pressure, it is fair to ask how organisms living in different physical environments accommodate these potentially disrupting effects on enzyme–substrate complex-formation. How, in other words, is the acetylcholinesterase binding function adjusted with respect to the physical environment?

In addressing ourselves to this problem, we began a series of experiments on purified electric-eel acetylcholinesterase, using a carbon substrate analogue and 'anionic-site' specific inhibitors (Hochachka, 1974).

The carbon analogue, 3,3-dimethylbutyl acetate,

$$CH_3\!-\!\underset{\underset{CH_3}{|}}{\overset{\overset{CH_3}{|}}{C}}\!-\!CH_2\!-\!CH_2\!-\!O\!-\!\overset{\overset{O}{\|}}{C}\!-\!CH_3$$

although similar in size and shape to the true substrate, differs in being uncharged. Binding at the anionic site is determined largely (perhaps solely) by hydrophobic forces, and therefore the compound provides a means for measuring the tempera-ture–pressure-dependence of this binding contribution. Kinetic studies of electric-eel acetylcholinesterase indicate that as temperature decreases from 35° to 2°C or when pressure rises to about 400 atm (40.5 MPa), the K_m for this substrate increases by about 10-fold. These results are consistent with predictions based on model compound studies, and with others encourage the view of low-tempera-ture high-pressure disruption of this contribution to binding (Fig. 1).

Such effects of temperature and pressure appear opposite in sign to the effects on the coulombic binding contribution. An indication of the latter arises from studies with inhibitors that compete with the substrate at the anionic site. For relatively simple substituted ammonium ions (such as dimethyl or ethyl ammon-ium ion), binding is determined largely by electrostatic interactions between enzyme and ligand (Froede & Wilson, 1972). The K_i for such a compound in the case of electric-eel acetylcholinesterase varies directly with temperature, but inversely with pressure, a result consistent with stabilization of these interactions by low temperature and high pressure (Hochachka, 1974).

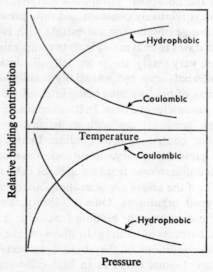

Fig. 1. *A diagrammatic summary of the effects of pressure and temperature on the hydrophobic versus coulombic contributions to binding substrate*

The figure is based on studies of pressure–temperature effects on purified electric-eel acetyl-cholinesterase with the uncharged carbon analogue 3,3-dimethylbutyl acetate as substrate and dimethyl ammonium ion as an inhibitor competitive with substrate for the anionic-binding site (Hochachka, 1974).

A picture thus emerges (Fig. 1) of an acetylcholinesterase binding function composed of two contributions: one, the coulombic contribution, favoured by low temperature and high pressure; the other, the hydrophobic contribution, disrupted by low temperature and high pressure. For this enzyme locus such a picture (1) identifies a strategic design rule (that the binding site be capable of interacting with both the polar and non-polar groups on the substrate) for minimum perturbation by environmental change, and (2) defines minimal design constraints (relative utility of each of the two potential binding contributions) for optimal acetylcholinesterase binding function in different physical environments.

This is an explanatory and potentially predictive picture. Without it, one would find it difficult to understand why acetylcholine should be any 'better' a transmitter substance than would, for example, 3,3-dimethylbutyl acetate, which would bind only by hydrophobic interactions (hence binding would be unusually sensitive to low temperature or to high pressure), or any 'better' a transmitter than would a simpler quaternary ammonium ion whose binding would be stabilized largely by the coulombic contribution (in which case binding would be unusually sensitive to high temperature and low pressure). With acetylcholine as substrate, if pressure is constant and only temperature varies, the disrupting effect of high temperature on the ionic contribution is nicely counterbalanced by the stabilizing effect on the hydrophobic contribution, and the reverse is the case at low temperatures. This physical situation is indeed the usual condition for terrestrial organisms and most aquatic ones; hence in these environments, an acetylcholinesterase binding site capable of interacting with both the charged and the non-polar groups on acetylcholine is clearly advantageous. If temperature is relatively constant and only pressure varies, the same counterbalancing advantage obtains for a substrate with two (hydrophobic and ionic) potential contributions to enzyme–substrate complexing. That is, whenever temperature or pressure vary singly, the potentially disrupting influences on one contribution are counterbalanced by potentially stabilizing influences on the other, the overall process of binding thus being internally compensated for with respect to one physical parameter at a time. In the deep sea, the physical conditions (low temperature, high pressure) are such as to favour the coulombic, but disrupt the hydrophobic, contribution to binding. In this environment, where both temperature and pressure co-vary, the situation is more complex, and our studies of brain acetylcholinesterase from an abyssal fish *Antimora rostrata* are therefore instructive, for if the above considerations are correct, two alternatives seem available to abyssal organisms. One possibility, and not a good one, would be to tolerate the disruption of binding functions; a better one would be to modify the functional aspects of binding. In *Antimora* the outcome of adapting to the extreme physical conditions of the abyss is an acetylcholinesterase that, compared with the homologous enzymes in high-cell-temperature organisms, displays a 'smaller' hydrophobic pocket in the substrate-binding site, playing a measurably diminished role in binding substrate. The consequent decrease in the contribution of $T\Delta S$ to the free energy of binding is closely compensated for by an increased enthalpic contribution, and the overall ΔG^0 of binding is therefore essentially unchanged (Hochachka *et al.*, 1975a). These design modifications constitute a simple yet fascinating example of biochemical 'optimization' leading to an acetylcholinesterase binding function that is fully compensated for with respect to the external temperature and pressure. [To an audience of biochemists, it may be worth stressing the magnitude of this optimization effect: what we are saying is that binding functions of *Antimora* acetylcholinesterase at low temperatures (2–3°C) and high pressures, up to several hundred atmospheres, are essentially equivalent to those of the ox brain enzyme homologue at relatively high temperatures (37°C) and 1 atm (101 kPa).]

This optimization appears not to be an isolated example, but rather to represent a general strategy of adaptation, and different versions of it arise over and over again with different enzymes. For example, in similar fashion we studied the pyruvate-binding site on M_4 lactate dehydrogenase using oxamate, an isosteric and isoelectronic substrate analogue, as a competitive inhibitor (Hochachka *et al.*, 1976). Here during ligand binding a negative charge must be neutralized, presumably by an arginine residue, arginine-171 (Holbrook *et al.*, 1975), a process that should be favoured by low temperatures and low pressures. The situation seems simple enough to be explained by a direct K_i–pressure and direct K_i–temperature relationship essentially identical for M_4 lactate dehydrogenases in general. What we, in fact, encountered is another example of biochemical optimization, with stringent selection for each enzyme to display a similar binding affinity in its particular physical environment. Thus, of all those animals studied, in all M_4 lactate dehydrogenases from 1 atm-adapted species, but not from midwater or abyssal fishes, the binding of oxamate is disrupted by high pressure. For the M_4 lactate dehydrogenases of '1 atm' species, the observed volume increase during charge neutralization is, in fact, close to the value expected from model compound studies (Hochachka, 1974). In sharp contrast, the volume change is essentially zero on forming the ternary complex with the enzyme from the abyssal or midwater fish. Since it is evident that carboxyl-charge neutralization does occur during ligand binding, we assume that the expected volume increase associated with charge neutralization must be masked by a concomitant volume change of similar magnitude but opposite sign. Where this might occur or by what mechanism, at the moment we do not know, except that it probably is remote from the substrate-binding site [see Somero & Low (1976), for further discussion of this area].

The situation with respect to temperature is perhaps even more dramatic. Although for any given lactate dehydrogenase, the $K_{i(oxamate)}$ increases in the thermal increment 15–45°C, the position of the K_i curve on the temperature axis correlates with the cell temperature at which the enzyme functions. Our first studies (Hochachka, 1975) compared M_4 lactate dehydrogenase homologues from the abyssal fish *Antimora*, adapted to function at about 2°C, and from the ox, a 37°C-adapted mammal. In addition, the same data are now available for a high-temperature marsupial (the possum), two monotremes (the platypus and the echnidna spiny anteater), and a reptile (the goanna). Although the K_i for any given enzyme can vary by 5–10-fold over the thermal increment 15–45°C, the K_i values at the average cell temperature of function are remarkably similar and do not vary from each other by more than about twofold! That is, at their respective biological temperatures, each lactate dehydrogenase–NADH binary complex binds oxamate and, by implication, substrate with an equal affinity. This rather dramatic illustration of optimization thus confirms that even although the overall features of homologous binding sites must be similar, a point emphasized and re-emphasized (probably over-emphasized) by studies of sequence and structure, enzyme-binding interactions can be strongly compensated for with respect to the physical environment. This outcome appears to depend on two general mechanisms, which can be viewed as constraints in the functional design of binding sites. Thus, one way to compensate ligand binding for temperature or pressure change is to modify the relative importance of different weak-bond contributions to stabilizing enzyme–ligand complex, a mechanism that may contribute to the adjustments we have noted in the binding functions of M_4 lactate dehydrogenases and acetylcholinesterases. A second way is to couple ligand binding with secondary reactions, spatially or temporally separated, which compensate for the enthalpy, entropy and volume changes occurring during substrate binding. Some such mechanism, for example, must be assumed to account for the pressure-insensitive binding of oxamate by the *Antimora* lactate dehydrogenase–NADH binary complex. In some cases, the enthalpy, entropy and volume changes associated with enzyme binding of allosteric modifiers (e.g. pyruvate kinase binding of fructose 1,6-diphosphate) may, in fact, serve to compensate for the same thermodynamic changes associated with substrate binding, in which instance the allosteric modifier would play a role in thermal and pressure compensation in addition to playing its usual role in regulation of enzyme activity. This possibility is currently being investigated in our laboratory. However it turns out, the above information indicates that fine evolutionary tuning of enzyme functional design allows homologues of the same enzyme to work efficiently despite widely disparate physical environments. At this time, similar 'optimization' of binding functions with respect to the physical environment are known for the NADH-binding sites of M_4 lactate dehydrogenases (Hochachka, 1975; Hochachka *et al.*, 1976), and octopine dehydrogenases (Luisi *et al.*, 1975), and for the substrate-binding sites of adenosine deaminases and pyruvate kinases (Guderley *et al.*, 1976). In each instance, although the structure of the enzyme binding site is not known, environmental selective forces acting on it are relatively well understood because of an appreciation of at least the major functional design features of that site. [In fact, in the case of dogfish M_4 lactate dehydrogenase, where 3-D structure is known, insight into such design features provokes a great deal of research into physical and chemical mechanisms by which they are achieved (see Holbrook *et al.*, 1975).]

Do the above considerations have a general validity? For this symposium I think that is a relevant question, for if they do, they imply that an understanding of biochemical systems in terms of design principles should lie at the heart of the problem of biochemical adaptation, and should add an important predictive element to studies in an area that to date has been largely descriptive (even if entirely fascinating!). How does one approach the problem? One way would be to do what a Martian would do if he landed on earth and wanted to decipher design principles of the automobile. The Martian would begin to take apart small cars, large cars, fast cars, slow cars, trucks, tractors and tanks; and the more vehicles he took apart and examined, the more precise would be his understanding of them. Biochemistry has clearly done exactly this with living systems, and indeed biochemistry textbooks are replete with generalized metabolic maps. Each of these is in effect a tidy summary of our current understanding of design principles, an example of which is offered in Fig. 2. This metabolic diagram is comparable in our analogy with a diagram outlining the general design principles of any motor vehicle, and would be equally valid for a tractor, a truck, a Porsche or equally valid for liver, muscle and gill. As such, it is very useful for understanding the rough outlines of metabolism, but clearly it describes nothing that would be instructive to a Martian wishing to know how to construct liver cell metabolism, how it would differ from that in muscle nor how it would vary from muscle type to muscle type. In our analogy, this level of generalization could not tell an engineer the design differences between, say, a fast and a slow automobile. But the basic idea, that design principles are as requisite as basic physical and chemical ones, is certainly inherent (though usually unstated) in summary diagrams of the type shown in Fig. 2. So what I would like to do with the remainder of this review, is to consider what we can currently assess of 'design rules' at these more integrated levels, using as a model system, the organization and control of muscle metabolism.

When faced with a similar problem, biologists often examine organisms living in environmental gradients; there they look for characteristics that seem to be sensitive to the gradient and spread along it. From such studies, it is known for example that teleosts living in environments of varying O_2 availability develop gill surface areas that are inversely proportional to O_2 availability. Otherwise,

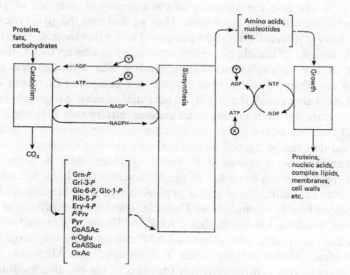

Fig. 2. *A block diagram summary of the three major functional 'units' of metabolism*

Taken from Atkinson (1977). Abbreviations: Grn-*P*, dihydroxyacetone phosphate; Gri-3-*P*, 3-phosphoglyceric acid; Glc-6-*P*, Glc-1-*P*, glucose 6-phosphate, glucose 1-phosphate; Rib-5-*P*, ribose 5-phosphate; Ery-4-*P*, erythrose 4-phosphate; *P*-Prv, phosphoenolpyruvate; Pyr, pyruvate; CoASAc, acetyl-CoA; α-Oglu, α-oxoglutarate; CoASSuc, succinyl-CoA; OxAc, oxaloacetate.

their morphology may be little changed, and from such simple observations it is clear to the biologist that gill surface area is an adaptable morphological characteristic. Similarly, we may be able to demonstrate which features of muscle metabolism are most sensitive to adaptive change by considering the adjustments made with respect to O_2 availability. Burst muscle work in vertebrates, since it is characterized by a rapid depletion of O_2, is a particularly well studied and instructive example of an energy metabolism designed for anaerobic function, and we shall begin our analysis with it.

Design of Vertebrate Muscle Metabolism for Burst Work

At the outset, it is worth emphasizing that animals follow two distinct strategies of adaptation to limited O_2 availability. One of these is basically compensatory in nature: in response to the temporary depletion of O_2, a good anaerobe activates anaerobic mechanisms of ATP production to correct for, or compensate for, the lack of O_2. In vertebrates, anaerobic metabolism is always glycogen based; the key requirements for maximizing its utility can be summarized as follows: (1) high glycolytic potential; (2) large percentage change from low- to high-activity states; (3) 'complete on–complete off' catalytic behaviour; (4) tolerance of high lactate accumulation.

How are these unique requirements achieved and organized into cellular energy metabolism? From a careful comparison of skeletal muscle with other tissues such as liver, it is evident that the simplest way of increasing the glycolytic capacity of a cell is to increase the amounts of the component enzymes of glycolysis maintained in the cell at any given time. Thus we find that the glycolytic enzymes in muscle occur in unusually high titre compared with other tissues. As an example, the specific activity of muscle phosphofructokinase can be up to 100-fold higher than that of liver phosphofructokinase, and similar although less dramatic differences are observed when other enzymes are compared (Scrutton & Utter, 1968). All else being equal, the glycolytic potential of muscle is greatly increased. Secondly, nature has built in muscle an enzyme battery that is strongly 'poised' for overall function in the glycolytic direction (i.e. for glycogen → lactate conversion) and that can be 'turned on and off' with extreme efficiency.

This characteristic is discussed in detail by Hochachka & Storey (1975); suffice it here to mention but a few of these properties (see Figs. 3a and 3b). For example, after the activation of glycogen phosphorylase and myofibrillar adenosine triphosphatase in normal skeletal muscle, rising concentrations of fructose 6-phosphate, fructose 1,6-diphosphate, ADP, AMP and P_i coupled with falling concentrations of creatine phosphate and ATP can lead to a very large 'flare-up' of phosphofructokinase activity when it is required (see Mansour, 1972). A central feature of this control system is that two of the positive modulators are products of the reaction; one of the positive modulators is, of course, a substrate. Taken together, their regulatory effects lead to an autocatalytic, exponential rate of change from low-activity states to high-activity states, a characteristic that often is not seen in other cells, and that helps to explain the speed with which muscle glycolysis is 'turned on'. Aside from this characteristic, other distinguishing

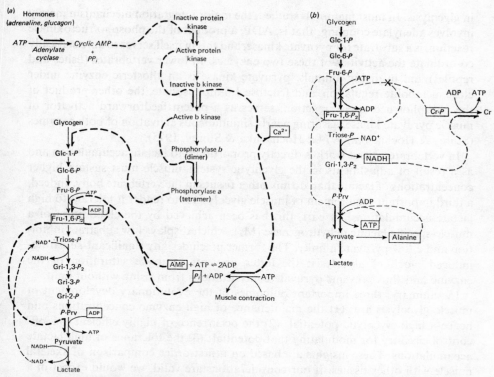

Fig. 3. *Metabolic maps showing currently held concepts of glycolytic activation in skeletal muscle of vertebrate animals (a) and control circuitry in heart glycolysis of the diving turtle (b)*

In (*a*) regulatory metabolites are connected up with the enzyme steps they activate by dashed lines; activation is indicated with a dark arrow (→). One aspect of this control diagram, fructose 1,6-diphosphate activation of pyruvate kinase, is thought to occur widely only among lower vertebrates; all other aspects are thought to be general, although certain characteristics are adjusted in diving vertebrates (see *b*). (*b*) shows that a major modification in the control set-up of the diving animal is the replacement of ATP inhibition with creatine phosphate inhibition of phosphofructokinase. Activation of creatine kinase by NADH is integrated with glycolytic activation. Fructose 1,6-diphosphate takes on a particularly pivotal role in that it (i) reverses creatine phosphate inhibition, (ii) activates its own production by phosphofructokinase and (iii) activates pyruvate kinase by a feedforward mechanism. Both (*a*) and (*b*) are from Hochachka & Storey (1975). Abbreviations: Fru-6-*P*, fructose 6-phosphate; Fru-1,6-P_2, fructose 1,6-diphosphate; Triose-*P*, triose phosphate; Gri-1,3-P_2, 1,3-diphosphoglycerate; Gri-3-*P*, 3-phosphoglycerate; Gri-2-*P*, 2-phosphoglycerate; *P*-Prv, phosphoenolpyruvate; Cr-*P*, creatine phosphate; Cr, creatine. Other abbreviations are as indicated in the legend to Fig. 2.

features of muscle phosphofructokinase compared with that in tissues such as the liver include a 'tighter' overall control by most organophosphate modulators and by citrate, but a much decreased sensitivity to ATP inhibition. Thus, for a given percentage change, muscle phosphofructokinase requires about one-tenth as much ADP, AMP or citrate as does the liver homologue; at the same time, it requires two to three times higher ATP concentrations to bring about the same percentage inhibition.

The question also arises of how the phosphofructokinase catalytic rate is integrated with the activity of pyruvate kinase, the next major regulatory enzyme in glycolysis. In most mammals studied, the major integration mechanism merely involves adenylate coupling; that is, ADP, a product of the phosphofructokinase reaction, is a substrate for pyruvate kinase, and this in itself serves to automatically co-ordinate the activities of these two enzymes. In lower vertebrates (fishes and reptiles) and in diving animals, pyruvate kinase is an allosteric enzyme under close metabolite regulation, and fructose 1,6-diphosphate, the other product of the phosphofructokinase reaction, serves as a potent feedforward activator of muscle pyruvate kinase, assuring nearly simultaneous activation of both enzymes (Storey & Hochachka, 1974a; Hochachka & Storey, 1975).

In vertebrates, burst work is directly proportional to lactate accumulation and as a result of adjustments in the glycolytic system, muscle must sustain higher concentrations of lactate than do any other tissues in the vertebrate body. Indeed, a third important requirement of muscle glycolysis is to render it tolerant to high lactate accumulations. In part, this has been achieved by the elaboration of a muscle-specific lactate dehydrogenase (M_4), which displays low substrate inhibition and a low pyruvate affinity. The former precludes any significant substrate-induced block of glycolysis; the latter prevents pyruvate saturation of the enzyme and thus prevents pyruvate concentrations from rising without limit.

In summary, three important outcomes of the evolutionary development of muscle glycolysis are: (1) the maintenance of high enzyme concentrations and hence a high glycolytic potential; (2) the occurrence of highly efficient 'on–off' control circuitry for modulating that potential; (3) the tolerance of high lactate accumulations. These insights are based on a first-order comparison of skeletal muscle with other tissues. If our considerations are valid, we would expect in a detailed study of skeletal muscle in superior anaerobes, that properties of the glycolytic system will be further adjusted. For this reason it is instructive to consider the metabolism of diving reptiles, birds and mammals, groups that include some of the best-known vertebrate anaerobes.

Muscle Glycolysis in Marine Mammals

The basic design of muscle glycolysis is so effective that when muscle of a diving mammal, such as the porpoise, is compared with an animal such as the common laboratory rat or man only a small number of modifications are observed, and these are all in effect elaborations on the theme already described. Thus with regard to muscle enzyme concentrations within the glycolytic chain, the major differences between a marine mammal and the laboratory rat are (a) increased levels of phosphoglucomutase, aldolase, α-glycerophosphate dehydrogenase and lactate dehydrogenase, and (b) decreased levels of pyruvate kinase. A third notable change is the presence of unusually high levels of fructose diphosphatase, a non-glycolytic enzyme nonetheless involved in glycolytic control. And fourthly, preliminary evidence suggests an unusually fast phosphorylase activation ($b \to a$ conversion) in diving animals [see Hochachka & Storey (1975) for a review].

On closer examination, one set of observations can be readily explained. Thus, because muscle in marine mammals is cut-off from blood circulation during diving, it must place a lesser reliance on blood glucose and a greater reliance on muscle glycogen than must that of a terrestrial mammal; this requirement in turn is reflected in a more sensitive phosphorylase control and in increased levels of phosphoglucomutase. Secondly, an active aldolase and fructose 1,6-diphosphatase contribute to the tight regulation of fructose 1,6-diphosphate concentrations; the latter is, of course, a key regulatory metabolite in the overall pathway. And thirdly, muscle cells in divers must be prevented from becoming highly reduced and this requirement is reflected in higher titres of lactate dehydrogenase and α-glycerophosphate dehydrogenase. These adjustments allow for higher lactate accumulations (in diving turtles values of over 100 mM-lactate can occur), which may lead to the evolution of key enzymes such as pyruvate kinase with lowered pH optima (pH 6.5 optimum in diving animals compared with pH 8.5 in the rat).

The Regulatory Nature of Muscle Pyruvate Kinase in Diving Mammals

An explanation for the comparatively low levels of pyruvate kinase within muscle of the porpoise had to await detailed kinetic studies of the enzyme (Storey & Hochachka, 1974b). Others had observed that, when divers are treated as a single group, the levels of pyruvate kinase in muscle correlate with the length of dive each species is capable of achieving and such a correlation between glycolytic capacity of the tissue and its pyruvate kinase titre, in fact, also holds for normal mammalian tissues (Simon et al., 1974). However, within diving animals as a group pyruvate kinase concentrations are lower than in the common laboratory rat. What diving animals seem to have done at this locus is to accept an overall decrease in the activity of pyruvate kinase in favour of designing an enzyme that is much more sensitive to metabolite regulation. Unlike the enzyme in terrestrial mammals, muscle pyruvate kinase in diving vertebrates is under tight feedback inhibition by creatine phosphate, ATP, alanine and citrate (Storey & Hochachka, 1974a,b) and under strong feedforward activation by fructose 1,6-diphosphate, which returns the pyruvate kinase maximum potential to the high range expected for a highly active glycolysis. Moreover, as with other regulatory pyruvate kinases, fructose 1,6-diphosphate not only directly activates the enzyme (by both K_m and $V_{max.}$ effects) but also strongly reverses the inhibitory effects of negative modulators. These regulatory characteristics are, in fact, commonly observed in muscle pyruvate kinases of lower animals, but appear to have been lost in most mammals. Diving animals have retained (or regained) this tight control over pyruvate kinase in part because of a high reliance on glycolysis during diving, but mainly because of important control requirements imposed on muscle at the end of the dive, i.e. during anaerobic–aerobic transition, when muscle metabolism 'switches' from glycogen to other fuels. These requirements arise from the fact that when the animal is not diving, the preferred fuel is fat, and standard R.Q. values are in the range of 0.7 for most marine mammals thus far studied (Matsuura & Whittow, 1973). In the rat, during anaerobic–aerobic transition, activation of the β-oxidation spiral leads to a momentary piling up of acetyl-CoA as oxaloacetate reserves for citrate synthesis are inadequate (Safer & Williamson, 1973). In diving animals such a limitation could be an unfortunate handicap. Thus the diving habit leads in muscle to another important metabolic

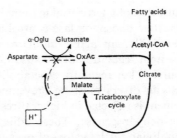

Fig. 4. *Favoured direction of aspartate aminotransferase function during anaerobic–aerobic transition in diving vertebrates such as the porpoise*

The aspartate affinity of the mitochondrial enzyme is about fivefold higher than that of the cytoplasmic isoenzyme, and it therefore competes effectively for intramitochondrial aspartate. Its K_m for oxaloacetate is about tenfold higher than intramitochondrial oxaloacetate concentrations, a factor that also favours function in the direction of oxaloacetate production. Moreover, malate concentrations are increasing at this time, owing to tricarboxylate-cycle activation, and malate serves as a potent inhibitor of the backward reaction. A low pH occurring at the end of the dive further potentiates the malate control at this locus. This scheme is taken from Hochachka & Storey (1975).

requirement: a source of oxaloacetate that can be 'turned on' particularly efficiently during fatty acid oxidation and tricarboxylate-cycle activation. The source of oxaloacetate is aspartate via the aspartate aminotransferase step. In marine mammals, such as the porpoise, the activity of this enzyme in muscle is up to 17 times higher than in terrestrial species (Owen & Hochachka, 1974) and its catalytic properties are consistent with oxaloacetate production in the mito-chondria during anaerobic–aerobic transition (Fig. 4). Glutamate, produced in the reaction, in turn transaminates with pyruvate to regenerate α-oxoglutarate; this process, catalysed by alanine aminotransferase, which also occurs in unusually high levels in muscle of marine mammals, leads to alanine accumulation (Hochachka *et al.*, 1975c). The total amount of alanine accumulated under such conditions is equal to the summed increase in concentration of all tricarboxylate-cycle intermediates (Safer & Williamson, 1973). That is a fundamental insight, for it emphasizes that alanine is singly perhaps the 'best' metabolite signal of the degree to which the tricarboxylate cycle is activated. It is therefore not surprising that alanine is such a good inhibitor of pyruvate kinase in diving vertebrates, for the greater the degree of tricarboxylate-cycle activation, the greater the degree to which pyruvate kinase is alanine-blocked and carbohydrate reserves are 'spared' for anaerobic excursions (Fig. 3). Thus the functional design of muscle pyruvate kinase in diving animals fits well the high glycolytic potential required during diving but also contributes to the solution of another design requirement of the diving habit, i.e. a metabolic organization that can oscillate efficiently between glycogen catabolism (during anaerobic portions of the dive) and fat catabolism (during aerobic work, such as surface swimming, migrations and so forth).

Anaerobic Glycolysis in Octopus Mantle Muscle

A comparable muscle metabolism, organized for burst-work situations, is found in the mantle muscle of the octopus (Fields *et al.*, 1976). As in the analogous

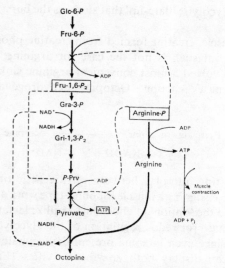

Fig. 5. *A tentative summary of control circuitry in glycolysis of octopus mantle*

Activation indicated by dark arrow; inhibition by dark cross. A central feature of this control circuitry is the inhibition by arginine phosphate of phosphofructokinase and pyruvate kinase. During burst muscle work, decreasing arginine phosphate concentrations are thought to lead to a de-inhibition of both these key steps in glycolysis. In addition, fructose 1,6-diphosphate serves as a feedforward activator of pyruvate kinase and would contribute to the activation of pyruvate kinase at an appropriate time. Arginine formed in the process serves as a co-substrate for octopine dehydrogenase, which takes on the role of lactate dehydrogenase in this tissue. It is evident that in burst work, the arginine+arginine phosphate pool could limit glycolysis at the octopine dehydrogenase step. Data from Fields *et al.* (1976). Gra-3-*P*, glyceraldehyde 3-phosphate.

vertebrate tissue, the octopus mantle muscle displays high levels of glycolytic enzymes and high concentrations of storage carbohydrate (and hence has a high glycolytic potential). Two sharply contrasting features of the octopus mantle, however, include a total absence of (*a*) lactate dehydrogenase and (*b*) creatine phosphate. Arginine phosphate takes over the role of creatine phosphate and octopine dehydrogenase takes over the role of lactate dehydrogenase (see Fig. 5). Arginine phosphate is indeed a common phosphagen in cephalopods (and other invertebrates) and occurs in concentrations of about 30–50 mM. In addition to its usual role in buffering oscillations in ATP concentration during muscle activation or recovery, arginine phosphate plays at least two important regulatory roles in being a potent negative modulator of mantle phosphofructokinase and pyruvate kinase (Guderley *et al.*, 1976). In the octopus, mantle-muscle pyruvate kinase is a closely regulated enzyme, under feedforward activation by fructose 1,6-diphosphate and feedback inhibition by arginine phosphate and ATP, both of which would contribute to glycolytic inhibition during energy-saturated periods. On initiation of burst work, falling arginine phosphate concentrations in concert with rising fructose 1,6-diphosphate concentrations satisfactorily account for a large pyruvate kinase activation at an appropriate time. Similarly, preliminary studies indicate that falling concentrations of arginine phosphate lead to de-inhibition of phosphofructokinase (K. B. Storey, personal communication), a process also contributing to the glycolytic 'flare-up' that sustains the burst work of an octopus attack or escape.

In mammalian muscle, creatine formed from creatine phosphate accumulates during burst work, but such is not the case for arginine in octopus mantle. Instead, nature has evolved a most convenient arginine sink in the form of the octopine dehydrogenase reaction. Octopine dehydrogenase, catalysing the following reaction,

$$\text{Pyruvate} + \text{arginine} \underset{\text{NADH} \qquad \text{NAD}^+}{\xrightarrow{\hspace{3cm}}} \text{octopine}$$

represents an interesting interplay between phylogenetic time and adaptation, for it is evident that in designing a muscle for burst-type work ancient cephalopods, at least in theory, had the 'option' of choosing to utilize lactate dehydrogenase or octopine dehydrogenase for redox regulation of the glycolytic pathway. I mean this fairly literally, since even in some present-day cephalopods a single tissue (e.g. squid brain) can display both enzyme activities (Fields *et al.*, 1976). Although interpreting the basis for the octopine dehydrogenase option taken by cephalopods is necessarily speculative, it appears to have been associated with the use of arginine phosphate as the muscle phosphagen. It will be evident (Fig. 5) that in a burst-type tissue, there will be a close correlation between the pool size of arginine and the physiological usefulness of octopine dehydrogenase, and indeed, it is also evident that glycolysis in the octopus mantle would be limited not only by the accumulation of end product but by the depletion of arginine phosphate. Should the ATP demands of muscle work surpass the capacity of the arginine phosphate pool plus the capacity of glycolysis to sustain a steady ATP supply, it is clear that the pathway would become limited by availability of arginine for the octopine dehydrogenase reaction, and thus by NADH accumulation. It is currently believed that this difficulty does not occur because of a large arginine plus arginine phosphate pool, and because octopine dehydrogenase occurs in octopus mantle at extremely high catalytic potential, at activities that at 25°C are as high as 600 units/g wet weight (Fields *et al.*, 1976). Porpoise muscle, by comparison, contains about 1200 units of lactate dehydrogenase activity/g at 37°C (Storey & Hochachka, 1974*b*). During burst work, both pyruvate and arginine would presumably be formed simultaneously, and both, of course, are substrates for octopine dehydrogenase. Substrate saturation for both substrates follows Michaelis–Menten kinetics and currently there are no known modulators, except for a regulatory effect involving the co-substrates, pyruvate and arginine; increasing availability of one substrate increases the affinity for the co-substrate. When both substrates increase in concentration simultaneously, the resulting saturation curve is clearly sigmoidal. Under normal conditions, therefore, a relatively small change in availability of substrates causes a relatively large change in reaction velocity. That is, the enzyme can be 'turned on or off' with high efficiency, and this is, of course, a hallmark of glycolysis in burst-work muscle.

Thus in two phylogenetically quite different groups of organisms the design of muscle metabolism for burst work shows some interesting similarities. In both instances, the basic strategy calls for a high power output (high yield of ATP/unit time) but low efficiency (low yield of ATP/mol of starting substrate). In both instances, the three key design rules are (1) a high potential achieved by high levels of glycolytic enzymes and storage substrates, (2) a tight control of that potential for allowing rapid transitions from low- to high-activity states and (3) a tolerance for high accumulations of end product (lactate or octopine). By these means the organism can compensate for the temporary depletion of O_2 in its 'burst-work' fibres by a large increase in the glycolytic production of ATP, and so can greatly extend the amount of work that can be performed under these conditions.

A point of emphasis is that glycolytic mechanisms are not well designed for long-term anaerobic capacity. When a prolonged tolerance to anoxia is, in fact, glycolytically based, as in the diving turtle *Psuedemys scripta* it is of use to aestivation and only under such conditions of decreased energy demands is it possible for the organism to survive rather indefinite (winter-long) time-periods without an electron-transport-mediated O_2 consumption. The limitation of this scheme arises from three key problems: (1) the metabolic cul de sac formed by either the lactate dehydrogenase- or the octopine dehydrogenase-type of reaction; (2) the low ATP yield/mol of substrate metabolized; (3) the need to pay back an O_2 debt. These are all readily soluble difficulties for organisms or tissues only temporarily deprived of O_2, but any or all of them would strongly select for alternative solutions in organisms invading anoxic environments on a sustained basis. Not surprisingly, therefore, good invertebrate anaerobes, those capable of facultative or even obligate anaerobiosis, have not become 'locked' into glycolysis in its simplest form as the sole device for anaerobic ATP formation.

Design of Muscle Metabolism for Sustained Anoxic Function

Many invertebrate groups have representatives capable of rather astonishing excursions into anoxic environments; of these, the best studied are the helminth parasites and the marine bivalves, such as the common oyster *Crassostrea*, the mussel *Mytilus* and the clam *Rangia*. Whereas all these marine bivalves seem capable of respiring when O_2 is available, they all also are capable of surprisingly long exposures to completely anoxic environments. The anaerobic metabolism of *Crassostrea gigas*, for example, can sustain the organism for a period of about 4 weeks at 25°C, a temperature approaching the upper biological limit for the species. Muscle tissue of this organism displays few mitochondria and its respiration rate is low even if O_2 is present. Whereas glycogen has long been known to occur in unusual abundance and to serve as an important anaerobic substrate, only recently has it become evident that during anoxia these organisms rely on the simultaneous mobilization of two energy sources, carbohydrates and amino acids. In consequence, they seldom produce lactate as a sole anaerobic end product; indeed, lactate is usually not produced at all because the enzyme for its formation is not present. Instead, a multiplicity of anaerobic end products are formed, the most important being metabolic CO_2, alanine and succinate. Other end products include volatile fatty acids such as acetate, propionate, isovalerate, isobutyrate and methylbutyrate (see de Zwann *et al.*, 1976; Saz,

1971; Hochachka *et al.*, 1973; for reviews see Stokes & Awapara, 1968). Although not all aspects of this metabolic organization are described, it is already clear that it allows these organisms to live in environments that would otherwise be unavailable to them and that underlying such exploitative adaptational strategies are mechanisms that couple other ATP-yielding reactions to those of classical glycoysis. As a result, the energetic efficiency (yield of ATP/mol of substrate) is improved, but power output (yield of ATP/unit time) is greatly decreased. That is why facultatively anaerobic invertebrates are typically sessile or at least sluggish and their muscle metabolic rate is extremely low. The key requirements for facultative anaerobiosis in marine invertebrates can be summarized as follows: (1) high potential for sustained anaerobic metabolism; (2) simultaneous catabolism of carbohydrates and amino acids; (3) multiple sources of high-energy phosphate compounds. We can illustrate how such requirements, in fact, are organized into the metabolic fabric of muscle in these interesting animals by considering the anaerobic metabolism of oyster muscle tissue; most of the data are from studies of oyster adductor, but some recent data also arise from studies of isolated oyster hearts.

A detailed consideration of the available data from the oyster adductor and oyster heart muscle suggests a metabolic organization (Fig. 6) with at least four linear paths:

1. Glucose → phosphoenolpyruvate → pyruvate → alanine
2. Glucose → phosphoenolpyruvate → oxaloacetate → malate → fumarate → succinate
3. Aspartate → oxaloacetate → malate → fumarate → succinate
4. Glutamate → α-oxoglutarate → succinyl-CoA → succinate

In this organization, the glycolytic pathway appears to be unchanged to the level of phosphoenolpyruvate (Fig. 6), but at the phosphoenolpyruvate branchpoint the situation becomes novel. Thus Collicutt (1975) found that in isolated oyster hearts made anoxic for up to 3 h, about 50% of glucose carbon flows to alanine, about 5% to succinate, and most of the rest appears in an unknown compound. Fields (1976) subsequently showed that the unknown compound is formed from pyruvate and alanine in an NADH-requiring enzyme reaction analogous to that catalysed by octopine dehydrogenase:

$$\text{Pyruvate} + \text{alanine} + \text{NADH} + \text{H}^+ \leftrightarrows \text{NAD}^+ + N\text{-(1-carboxyethyl)alanine}$$

This enzyme occurs in highest titre in adductor muscle, at intermediate activities in heart, and at lowest activities in the mantle. The product of the reaction is thought to accumulate during anoxia in a manner analogous to lactate in mammalian muscle.

In addition, in the oyster heart, aspartate concentrations are high, about 15 mM (Collicutt, 1975), and these are depleted during anoxia. Most (about 70%) of the aspartate carbon flows to malate and succinate; some (about 17%) accumulates as alanine. Currently, it appears that the oxaloacetate formed from either glucose or from aspartate behaves as a single metabolic pool, and it will be evident that the major fate of oxaloacetate is to be reduced to malate by cytoplasmic malate dehydrogenase. By comparison with other enzymes in oyster muscle, malate dehydrogenase occurs in high titre (about 100 units/g wet weight

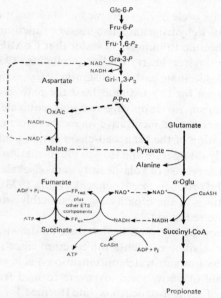

Fig. 6. *A tentative metabolic map of anaerobic metabolism in muscle of facultatively anaerobic marine bivalves, such as the oyster*

It will be evident that at least four linear paths are potentially operative. They are so arranged as to lead to the simultaneous mobilization of carbohydrate and the free amino acid pool (via aspartate and glutamate). In obligate anaerobes, such as *Ascaris*, the pyruvate kinase arm of the phosphoenolpyruvate branchpoint is deleted, and all glucose carbon flows to oxaloacetate and ultimately succinate (or propionate). In the oyster heart, in contrast, the pyruvate kinase arm appears to be favoured, whereas in adductor muscle in prolonged anoxia, although the pyruvate kinase activity is potentially high, it is thought to be fully inhibited and all glucose is again thought to flow towards succinate. The data for this scheme were taken from Hochachka *et al.* (1973) and from Collicutt *et al.* (1976). FP$_{red.}$, FP$_{ox.}$, reduced and oxidized flavoprotein respectively; ETS, electron-transfer system.

of adductor muscle at 25°C) and, together with the above pyruvate–alanine-requiring reaction, it takes on the function of lactate dehydrogenase in regenerating NAD$^+$ for glycolysis. Not surprisingly, lactate dehydrogenase is absent. Unlike the lactate formed in vertebrate tissues, the malate formed in oyster muscle does not accumulate; rather it is dehydrated to form fumarate, which in turn is reduced to succinate in a reaction catalysed by 'fumarate reductase' and yielding 1 mol of ATP/mol of succinate formed. That is, fumarate in these tissues is thought to behave as a terminal electron acceptor (see De Zoeten & Tipker, 1969).

To this point, it is important to emphasize, that the system is in redox balance in the cytoplasm, but intramitochondrially some oxidation reaction (or reactions) must be coupled with fumarate reduction. One candidate postulated for this job is the mobilization of glutamate to α-oxoglutarate; the latter as substrate for α-oxoglutarate dehydrogenase is converted into succinyl-CoA and sets the stage for a substrate-level phosphorylation in a reaction catalysed by succinic thiokinase

$$\text{Succinyl-CoA} \longrightarrow \text{succinate} + \text{CoA}$$

$$\text{GDP} + \text{P}_i \qquad \text{GTP}$$
$$(\text{ADP} + \text{P}_i) \qquad (\text{ATP})$$

As in the tricarboxylate cycle during aerobic metabolism, it is evident that in this scheme α-oxoglutarate dehydrogenase also must be functionally coupled in a 1:1 activity ratio with succinic thiokinase in order that CoASH be neither depleted nor accumulated. In oyster heart, studies (Collicutt, 1975) have established that the glutamate → succinate path is substantially less active than the aspartate → succinate one. Hence for this tissue at least the pathway cannot represent a single unique mechanism for supplying reducing equivalents to fumarate reductase. In other invertebrate facultative anaerobes, formally analogous pathways for the mobilization of the branched-chain amino acids (leucine, isoleucine and valine) apparently do operate and lead to the formation of 1 mol of ATP/mol of amino acid, with the release of volatile fatty acids (isovalerate, isobutyrate and methylbutyrate) as end products (see Hochachka et al., 1973). The oxocarboxylate dehydrogenase reactions in these pathways presumably would also form redox couples with fumarate reduction.

A close examination of Fig. 6 indicates that, as drawn, this scheme (which ignores the pyruvate–alanine condensation reaction since its activity initially is presumably limited by low substrate concentrations) is in redox balance, if, and only if, for each 2 mol of phosphoenolpyruvate formed from glucose, 2 mol of aspartate and 2 mol of some α-oxocarboxylate (formed for example from glutamate, or one of the branched-chain amino acids) are mobilized simultaneously. Indeed, some such general coupling seems ensured by at least two mechanisms: (1) by the redox couples formed between malate dehydrogenase and triose phosphate dehydrogenase and between fumarate reduction and α-oxocarboxylate dehydrogenases, and (2) by alanine aminotransferase. The latter, which catalyses the reaction,

$$\text{Pyruvate} + \text{glutamate} \rightarrow \alpha\text{-oxoglutarate} + \text{alanine}$$

also contributes to the coupling between carbohydrate and amino acid metabolism, since one substrate (pyruvate) is derived, at least in part, from carbohydrate, whereas the co-substrate (glutamate) is derived from the free amino acid pool. Furthermore, this coupling process explains one function for alanine as a major anaerobic end product in these organisms. Since the reaction is thermodynamically fully reversible, the kinetic properties of the enzyme in the oyster have been modified so as to favour a unidirectional catalysis (Mustafa, 1975).

Energy Yield in Anoxic Oyster Muscle

A comparison of the energy yield in oyster muscle relative to that of classical glycolysis is difficult because different substrates are used and different end products accumulate. Some of the substrates, such as aspartate, are not fermented by vertebrate muscle; propionate and other volatile fatty acids are never formed as anaerobic end products in vertebrates. However, when considering the relative energetic efficiency in these organisms, it is useful to recall that in obligate anaerobes (such as the well studied Ascaris lumbricoides), pyruvate kinase is often not present (Saz, 1971). Deletion of this arm of the phosphoenolpyruvate branchpoint decreases glucose fermentation to a linear path, either glucose 6-phosphate → succinate, or in some cases, glucose 6-phosphate → propionate; in the first case, the energy yield is 5 mol of ATP/mol of glucose 6-phosphate, and in the second case it probably is 7 mol of ATP/mol of glucose 6-phosphate. That is, the trend in these organisms is for maximizing energetic efficiency. Power output (yield of ATP/unit time) on the other hand, is probably fairly low. Isolated anoxic oyster hearts, for example, convert only about 1 μmol of glucose into alanine (with the accompanying formation of aspartate and succinate)/s per g wet weight, equivalent to less than 1 μmol of ATP/s per g. By comparison, trout muscle can convert 40 μmol of glucose into lactate/s per g during initial phases of burst work (Stevens & Black, 1966); this is equivalent to about 13 μmol of ATP/s per g, a value about an order of magnitude higher than that observed in the oyster.

In summary, it appears that invertebrate facultative anaerobes, such as the oyster, utilize an exploitative strategy of anoxia adaptation allowing them to invade anoxic environments and sustain anoxic local conditions for time-periods that as a fraction of their life-time can be viewed as indefinite. The primary rule of design is an energy metabolism of relatively high efficiency (relatively high ATP yield/mol of substrates) but low power output (low ATP yield/unit time), and these organisms are characteristically sessile or sluggish. Other design features of this metabolism include: (1) high concentrations of multiple substrates (mainly carbohydrates and free amino acids), but relatively low levels of enzymes (allowing an efficient but low rate of fermentation); (2) close regulation of the kind of enzyme present (e.g. phosphoenolpyruvate carboxykinase compared with pyruvate kinase, no lactate dehydrogenase or octopine dehydrogenase, a regulatory alanine aminotransferase, a 'fumarate reductase' etc.); (3) the use of mitochondrial metabolism for anaerobic purposes, a characteristic that singly, most strongly contrasts this metabolic organization with that of highly O_2-dependent systems.

An interesting example of a highly O_2-dependent muscle metabolism arises from our recent studies of a fast swimming predaceous squid. *Symplectoteuthis oulaniensis* is a vertical migrant, following its prey food source through the water column in a diurnal manner. At least three activity levels and hence metabolic states are clearly discernible. The lowest or basal metabolic rate is that required to sustain respiratory movements of the mantle, ion gradients, position in water column and so forth. An intermediate activity level is observed at night in areas such as the deep waters off the Kona coast of Hawaii, where 'packs' of this squid can be encountered in fairly high-velocity swimming, the basic hunting patterns of successful pelagic predators. Although critical data on how long this kind of performance can be sustained are lacking, the impression gained from visual observations is that it is a true steady-state muscle work that can be continued indefinitely. Finally, a third activity level is observed during attack on prey organisms (also presumably during escape from predators), and this swimming almost always involves violent bursts of speed. From available data it appears that a well developed high-pressure circulation system allows all three levels of muscle work to be supported by a vigorous energy metabolism that is probably obligatorily aerobic.

Design of a Highly Aerobic Metabolism in Squid Mantle

In mammals and other vertebrates, sustained aerobic muscle work as a rule is based on fat as the primary energy source, whereas, as we have seen, burst work is powered by carbohydrate fermentation. In some vertebrates (fish in particular) different muscle masses (red compared with white muscle) are usually involved in these different performance specializations, but in squid, this complication does not arise. *S. oulaniensis*, like other squids, depends on a jet-propulsion swimming mechanism both for sustained and burst activity and both kinds of activities are powered by a mantle muscle, which is structurally quite homogeneous (Moon & Hulbert, 1975). Fat content is less than 1.5% on a wet weight basis, hence fat is a trivial source of energy. The most important source of energy appears to be carbohydrate, although under some circumstances it is clear that proline and perhaps other amino acids can be utilized for catabolic purposes (Hochachka *et al.*, 1975d).

As far as we can tell, the pathway of glycolysis in squid mantle resembles the triose phosphate level schemes already described (Fig. 3). But at the level of dihydroxyacetone phosphate, the presence of extremely high activities (about 250 units/min per g at 25°C) of α-glycerophosphate dehydrogenase sets the stage for a serious potential drain of carbon from mainline glycolysis (Fig. 7). This

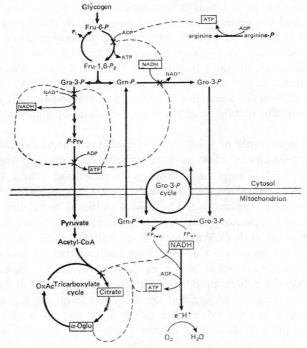

Fig. 7. *Metabolic map and known control circuitry for squid mantle-muscle energy metabolism*

Boxed metabolites are established modulators; dark crosses indicate reaction steps sensitive to inhibition; dark arrows, steps sensitive to activation. The α-glycerophosphate cycle is thought to be central both to the organization and to the control of this metabolism; this is discussed fully in the text. The scheme is taken from Hochachka *et al.* (1975d). Gro-3-*P*, 3-glycerophosphate.

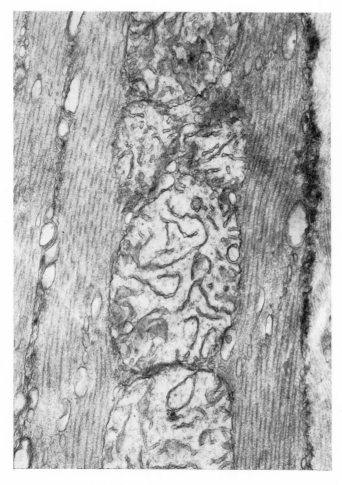

EXPLANATION OF PLATE I

Electron micrograph of a longitudinal view of squid mantle muscle showing the massive inner mitochondrial core, the abundant (if disrupted) intramitochondrial cristae and the dark glycogen-like granules along the boundary edge (in the upper right of the micrograph)

Magnification: ×20000. Taken from Hochachka *et al.* (1975*d*).

P. W. HOCHACHKA

potential difficulty is circumvented by the presence of an α-glycerophosphate cycle (discussed below) so that at steady state dihydroxyacetone phosphate in fact is neither depleted nor accumulated. Hence, despite this potential drain, all the glucose carbon apparently flows to pyruvate, which can then enter the mitochondria for complete oxidation. Proline, on the other hand, is thought to enter mitochondrial metabolism in the Krebs cycle at the level of α-oxoglutarate [see Sacktor (1976) for further discussion of this pathway]. Given this overall metabolic system, how is it designed to fit the high energy consumption that typifies this species' lifestyle?

As in our discussion of burst work, it will be readily appreciated that perhaps the simplest way to increase the aerobic potential of such a muscle metabolism is to increase the amount of requisite enzymic machinery, while simultaneously decreasing the activity of enzymes with anaerobic functions. Nature in fact utilizes this simplest strategy. Thus key enzymes for anaerobic glycolysis occur at low activities compared with 'burst-work' muscles, whereas those of aerobic glycolysis (phosphofructokinase, pyruvate kinase, α-glycerophosphate dehydrogenase and so forth) occur at high levels, as do the enzymes of mitochondrial metabolism, such as citrate synthase, malate dehydrogenase and α-glycerophosphate oxidase (Hochachka et al., 1975d).

Electron microscopy of squid mantle reveals muscle cells that are abundantly packed with mitochondria. The mitochondria are large and well endowed with cristae. In cross-section, each bundle of myofibrils surrounds an internal mass of mitochondria, the latter in effect filling the 'hole' of a doughnut formed by the myofibrils (Plate 1). In S. oulaniensis, over 50% of the muscle cell volume can be occupied by mitochondria (Moon & Hulbert, 1975), whereas in muscle of other cephalopods mitochondrial volume is as low as 10–15% (J. Arnold, personal communication).

Although all the above observations are wholly consistent with the concept of a highly aerobic metabolism in squid mantle, of themselves they give us little insight into how this metabolism correlates with the known swimming behaviours observed in the intact animal. What seems to be required here are activation mechanisms that can be set at, at least, two general levels: (1) at the level required to power steady-state swimming, the pattern used during hunting, and (2) at the level required to power the burst swimming used during prey capture. The first requirement in fact is common to active muscles in vertebrates and in insects. In both of the latter examples, glycolytic and tricarboxylate-cycle activation is regulated by the adenylates and by key metabolites within the pathways, each acting as signals for 'turning-on' specific regulatory enzymes. Similar control features are of course found in the squid mantle. Thus phosphofructokinase, an established control site in glycolysis, is under feedback inhibitory control by citrate but under positve modulation by P_i and particularly by AMP (Storey & Hochachka, 1975a). In other systems, the enzyme is also strongly inhibited by ATP, but in the squid the role of ATP seems to be taken over by NADH, the significance of which we shall discuss later. Pyruvate kinase, another potential control point in squid mantle glycolysis, is also a regulatory enzyme sensitive to inhibition by ATP, citrate and NADH; of these, the last appears by far the most effective (Storey & Hochachka, 1975b). Similarly, control of carbon

2

entry into the tricarboxylate cycle can be readily achieved through modulator influences on squid mantle citrate synthase; the enzyme is inhibited by ATP, citrate and NADH. ATP is competitive with respect to acetyl-CoA and citrate competes for the oxaloacetate site. NADH is again by far the most potent inhibitor. Concentrations of 0.01 mM cause a drastic 20-fold increase in the K_m for oxalo-acetate. This effect can be partially reversed by AMP, and fully reversed at high levels of oxaloacetate (Hochachka et al., 1975b). From these studies it is evident that during the transition from basal metabolism to steady-state swimming, momentary decreases in NADH, ATP and citrate concentrations, concomitant with rising concentrations of AMP and P_i could serve to activate aerobic glycolysis and the Krebs cycle. Because of an efficient mechanism for transferring reducing equivalents from cytosol into mitochondria, NADH regulation is particularly useful in co-ordinating the activation of glycolysis and the tricarboxylate cycle. It is believed that proline, which is stored at over 100 mM concentrations, is mobilized at the same time by mechanisms similar to those observed in insect flight muscle (Sacktor, 1970), and that it augments the pool of various Krebs-cycle intermediates, particularly the supply of oxaloacetate. By this mechanism, oxaloacetate is made available as substrate for citrate synthase at the same time as acetyl-CoA is being generated from glucose, an obviously advantageous arrangement.

Finally, and a point that I do not have time to properly summarize here, many of the above features of squid mantle metabolism, particularly many of the controlling interactions, are well compensated for with respect to temperature and pressure. That is an unusually important aspect of enzymic adaptation in S. oulaniensis, since it is thought to undergo daily migrations through substantial temperature and pressure gradients in the marine water column.

These kinds of mechanisms (and there undoubtedly will be more of them) would seem adequate to account for the transition from basal metabolic rates to those characteristic of the hunting pattern of swimming, wherever this should occur in the water column. But what of the next transition, the transition that powers the attack or escape bursts of swimming? In vertebrates, this kind of transition seems always to be powered by anaerobic glycolysis, and it behoves us to investigate this possibility first. It turns out that the anaerobic capacity of the mantle muscle in S. oulaniensis is low. Enzymes such as lactate dehydrogenase, alanine aminotransferase, phosphoenolpyruvate carboxykinase and 'malic' enzyme, which serve anaerobic purposes in other organisms, occur in low titre or are absent. Octopine dehydrogenase, although present, occurs at specific activities that are only about one-fifth of those of cytoplasmic α-glycerophosphate dehydrogenase. It has low affinities for both pyruvate (K_m of about 2 mM) and arginine (K_m about 12 mM). In exhausted muscle, the highest concentrations of pyruvate do not surpass about 0.7 mM, whereas arginine concentrations are about 5 mM. Thus the potential usefulness of the octopine dehydrogenase reaction seems greatly limited, except perhaps under quite unusual and unexpected conditions with pyruvate accumulating essentially without limit. Even then, it would be limited by a 5 mM-arginine pool. By comparison, if the mantle is forced into anoxia (as on exposure to air), α-glycerophosphate can accumulate to levels as high as about 20 mM. Sacktor (1970) has pointed out that energetically this situation is catastrophic because it greatly decreases the ATP yield (from a minimum of 0 mol of ATP/mol of glucose to 1 mol of ATP/mol of glucose 6-phosphate formed from polysaccharide). For these reasons, the possibility that this squid's highest swimming velocities are powered by anaerobic glycolysis does not seem too likely. We would be less hasty in this rejection if we did not have an obvious alternative: activation of an aerobic mechanism, i.e. the α-glycerophosphate cycle.

The cytoplasmic arm of the α-glycerophosphate cycle begins with the reaction

catalysed by α-glycerophosphate dehydrogenase. In squid muscle and bee flight muscle, the enzyme occurs at about 100-fold higher activities than it does in the oyster adductor and about 10-fold higher than in the octopus mantle. In the latter two muscles, cytoplasmic α-glycerophosphate dehydrogenase presumably plays an anaerobic role (emergency mechanism for reoxidizing NADH for glycolysis) but in the former, the enzyme has an aerobic function: formation of α-glycerophosphate for further metabolism in the mitochondria. Hence, a comparison of the kinetic properties of the homologous enzyme in the oyster, the squid and the bee seemed imperative (Storey & Hochachka, 1975c). Two observations are particularly instructive: (1) the squid and bee enzymes are both potently inhibited by ATP, the K_i values being 0.6 and 2.0 mM respectively, whereas the oyster enzyme is not affected by ATP at concentrations as high as 10 mM; (2) the squid and the bee enzymes both display low affinities for dihydroxyacetone phosphate, but high affinities for NADH, whereas the reverse is the case for the oyster adductor enzyme (Storey & Hochachka, 1975c). This means that the squid enzyme is designed to become most effective whenever dihydroxyacetone phosphate concentrations rise, provided that ATP concentrations are decreased. We envisage that the most likely time for this to occur is during the burst swimming of attack or escape. At such time, the sudden rise in ATP requirements for muscle contraction may momentarily lead to an unstable situation, with ATP demands exceeding supply capacities. These are necessary pre-conditions for maximal sparking of cytoplasmic α-glycerophosphate dehydrogenase and the α-glycerophosphate cycle. The consequent α-glycerophosphate dehydrogenase-mediated fall in NADH would automatically remove inhibition by at least three key glycolytic loci, phosphofructokinase, glyceraldehyde 3-phosphate dehydrogenase and pyruvate kinase, whereas the activation of mitochondrial respiration (mediated by α-glycerophosphate and increased ADP availability) would decrease intramitochondrial NADH, thus releasing citrate synthase from inhibition in a co-ordinated manner and at a physiologically appropriate time. One can readily appreciate that as soon as the energy status returns to its preburst value the reverse of these events (initiated by ATP inhibition of cytoplasmic α-glycerophosphate dehydrogenase) would again dampen the α-glycerophosphate cycle; the consequent readjustment in the NADH/NAD⁺ ratio would effectively slow down glycolysis and the tricarboxylate cycle. Perhaps for these reasons, the squid does not appear to be able to sustain the burst swimming pattern indefinitely, and, in the field, one observes that pursuit is abandoned if the squid fails to capture its prey after several consecutive attacks. (The most successful evasion pattern of prey fishes that I have observed almost always involved leaping out of the water, apparently leaving the squid mildly confused and still hungry!)

Finally, it will be readily recognized that the organization of metabolism in squid mantle leads to the complete oxidation of carbohydrate, and the yield of ATP/mol of glucose 6-phosphate is therefore high (36 mol). There is also the possibility that malate or oxaloacetate, formed from proline, can be decarboxylated to pyruvate for further and complete oxidation. In that event, the squid mantle gains also an additional 21 mol of ATP/mol of proline. However one estimates energetic efficiency, the value is a high one for squid mantle. Moreover, the power output is undoubtedly high. Though we have no ready measurement of it in squid mantle, the yield of ATP/unit time in insect flight muscle is probably one of the highest known in nature, and I would guess that the yield in mantle muscle of *S. oulaniensis* is in a similar range.

In summary then, the squid *S. oulaniensis* takes maximum advantage of O_2 as a terminal electron acceptor and through evolution has developed a mantle-muscle metabolism that is probably obligatorily aerobic. A number of fundamental design rules for this kind of metabolism are (1) a high aerobic potential (achieved by high enzyme titres, a high mitochondrial mass, a high respiratory rate primed by the α-glycerophosphate cycle); (2) an effective tricarboxylate-cycle activation mechanism (achieved by the utilization of two substrates, glucose and proline, the first generating acetyl-CoA, the second simultaneously augmenting the pool size of Krebs-cycle intermediates and oxaloacetate in particular); (3) an efficient mechanism for co-ordinating cytoplasmic and mitochondrial metabolism (achieved by linking control of both spans of metabolism to the redox potential as well as to the energy status); (4) an efficient means for transfer of reducing equivalents into the mitochondria (achieved by the α-glycerophosphate cycle); (5) a mechanism for 'square-wave' step up of aerobic metabolic rate for short emergency bursts of work (achieved by coupling the α-glycerophosphate cycle to the energy status of the cell through ATP inhibition and designing a cytoplasmic α-glycerophosphate dehydrogenase with appropriate kinetic properties); (6) mechanisms for temperature and pressure compensation of many of the above processes, particularly the control interactions, to allow migration through thermal and pressure increments in the marine water column (achieved at least in part by modifications of enzyme–ligand-binding interactions).

The resulting metabolic organization fits well the organism's lifestyle and environment and is remarkably similar to that already described for the flight muscle of some insects (Sacktor, 1970). In both these phylogenetically quite distinct groups, the overall strategy of design is a muscle metabolism of high power output (high ATP yield/unit time) and high energetic efficiency (high ATP yield/mol of substrate consumed), power output in both perhaps being the highest known in nature.

To recapitulate, we have seen three examples of designs for muscle metabolism, each apparently well suited for its particular environment. Unlike some parameters (such as the ratio of leg bone diameter/body weight or the affinity of a specific enzyme for its substrate), we have no ready measure of the degree to which a given metabolic organization has been optimized, but that each design is a compromise outcome is clearly evident. In burst-work metabolism, for example, there is an obvious compromise of long-term anaerobic capacity. Moreover, energetic efficiency is compromised for a high if anaerobic power output. In each species, the higher the demand for burst muscle work, the higher the ratio of glycolytically generated power output/energetic efficiency. As a matter of fact, energetic efficiency in this case is unchanging at 3 mol of ATP/mol of glucose 6-phosphate. Hence, as glycolytic potential rises (in response to increased need for burst speed), this ratio rises, the design clearly being optimized to meet the requirements of the whole organism for burst work (during activities such as running, flying and swimming). In the facultative anaerobe, the reverse compromise (sustained anoxia in exchange for high mobility) is made and the trend is to maximize energetic efficiency but to minimize power output; this design also seems to be in the process of optimization, for the further along the aerobic–anaerobic spectrum one goes, the more obvious is the trend. It is indeed a useful strategy and allows invertebrate anaerobes to exploit more fully anoxic environments for the lower the ratio, power output/energetic efficiency, the longer the organism can be sustained on any given amount of substrate. Finally, it might appear that the obligate-aerobic design gives the organism the best of all possible worlds (high power output and high energetic efficiency); it clearly is an optimization of muscle metabolism for organisms that have either an unlimited environmental O_2 supply or an essentially unlimitable O_2 delivery system or both. But these advantages are achieved at the cost of two disadvantages: first, the disadvantage of essentially zero anaerobic capacity (a highly compromising situation should the organism run into either a permanent or even a temporary O_2 depletion!), and secondly, in the case of certain aquatic forms (such as tuna and probably the fast swimming squid), the energy-cost disadvantage of having to be continuously moving through the medium in order to adequately oxygenate their gills and hence their muscles.

Whereas we can appreciate the biological significance of these different muscle metabolisms, it is important to emphasize that no amount of knowledge of the chemistry of carbohydrates, amino acids and fats or enzymes would ever allow us to predict the kind of metabolic organization that may arise in any given environment. If we ever can make such general predictions (and I hope that we may be able to do so), they will have to be based on higher-level design principles. That is why I have intentionally drawn on examples of muscle metabolism (largely carbohydrate-based) that span a range from obligate-anaerobic to obligate-aerobic organization, for by this analysis I hoped (1) to gain insight into the 'general design principles' of muscle metabolism, and in particular, (2) to assess which characteristics are adaptable and which are relatively conservative. (To return to our initial Martian analogy, what I have been doing is taking apart tractors, Fords and Alpha Romeos to see which features are common to all motor vehicles and which are adaptable for specific kinds of jobs and specific kinds of environments.) If there is a single certainty arising from this analysis, it is that in moving along the anaerobic–aerobic spectrum, not all characteristics of muscle metabolism are changed; rather, certain kinds of properties are changed time and time again, in different combinations and permutations. Others are relatively or highly conservative. It therefore is convenient, in conclusion, to consider the overall design of muscle metabolism (*a*) in terms of those characteristics that are invariant in muscle cells and (*b*) in terms of those characteristics that, although variant, are nonetheless essential.

Invariant Design Features of Muscle Metabolism

In all muscle cell energy metabolism, provision is made for at least six invariant features: (1) for redox stability (by a general 1:1 coupling between oxidative and reductive reactions under both anaerobic and aerobic conditions); (2) for CoASH cycling (by a 1:1 coupling between CoA-utilizing and CoA-releasing reactions under both anaerobic and aerobic conditions); (3) for adenylate coupling between key steps within any given pathway, between different pathways and between metabolism and muscle contraction; (4) for phosphagen-based buffering of changes in ATP concentrations during muscle work and during recovery; (5) for ion-dependent coupling between cell membrane depolarization and muscle activation. In addition, (6) the general outlines of control circuitry (i.e. coarse controls) are remarkably similar. Thus the same control points (such as glycogen phosphorylase, phosphofructokinase, pyruvate kinase, pyruvate dehydrogenase, citrate synthase, isocitrate dehydrogenase, α-oxoglutarate dehydrogenase, to name a few) appear over and over again no matter what the muscle type, no matter what the species. Not only are the same regulatory enzymes involved, some aspects of control such as the roles of energy charge and the adenylate pools, are also remarkably conservative.

These several features of energy metabolism in muscle (there may be more but these illustrate the point) are so well known to biochemists that they are only mentioned here for the sake of completeness and to emphasize that they appear invariant. Any mutations strongly modifying these characteristics would probably be lethal; thus such mutations are not the stuff of biochemical evolution and adaptation.

Variant but Essential Design Features of Muscle Metabolism

A second category of design rules for muscle metabolism (Table 1) appear much more flexible than those discussed above; they are variant in the sense that in any given subcategory not all possibilities need arise in evolution. Clearly, some alternatives are better than others for particular purposes and particular environments, and it is these, of course, that make up the material of biochemical adaptation. Thus, whereas an appropriate enzymic machinery, for example, is an absolute requisite of any muscle metabolism, the amounts and kinds of enzymes present are variant and correlate with the lifestyle and environment of the organ-

Table 1. *Muscle metabolism*

Variant but essential design features. 1. Enzyme amount (high amount, high potential). 2. Enzyme kind (defined in terms of reaction catalysed and regulatory properties). 3. Substrate kind and amount. 4. Fine-control circuitry. 5. Compensation mechanisms with respect to environmental parameters such as pressure, temperature, salinity, H^+, CO_2, O_2. 6. Anaerobic/aerobic blend.

ism. As we have seen, the simplest way to alter the overall catalytic potential of any given metabolic pathway is to alter the amount of enzymic machinery available at any given time. A high glycolytic potential can then be simply realized by increasing the steady-state amounts of glycolytic enzymes, as indeed appears to occur in tissues of diving vertebrates, for example (Hochachka & Storey, 1975). A high potential for fat catabolism can be similarly achieved by increasing the concentration of the enzyme machinery of the β-oxidation spiral, a situation that for economy purposes, might be correlated with a relative decrease in the amounts of glycolytic enzymes. Such a mutual, though not total, exclusion in fact appears to typify the metabolism of aerobic mammalian muscle fibres such as are found in the heart, the diaphragm, in fact in red-muscle fibres in general (Drummond, 1971). Where an anaerobic pathway and an aerobic one are both needed, as in the muscles of diving mammals, special control channels may arise to co-ordinate their relative activities. As we briefly illustrated in our example of muscle pyruvate kinase from diving animals, these design modifications almost always involve alterations in the catalytic and regulatory properties of key enzymes in one or the other pathway or in both. Hence, it is important to realize that we define 'kind' of enzyme not only in terms of the reaction it catalyses, but also by its regulatory and catalytic properties, by its position in a general metabolic 'field'.

Implicit in the above comments is the assumption that although some sort of substrate must be available for muscle metabolism, organisms appear to have at their disposal at least three great categories of substrate source, carbohydrates, amino acids and fats. Usually, of course, a combination of two or all three categories is used, with burst-work muscles (utilizing only carbohydrate) being a notable exception. The specific species of carbohydrate, amino acid or fatty acid of course varies as well. Depending on the substrate used, the end products released from either anaerobic or aerobic metabolism will also constitute a variant design feature of the system. Furthermore, just as enzyme amount is variant, so also is the amount of storage substrate maintained by any given type of muscle, and it is a common strategy indeed to boost the potential of a pathway like glycolysis by storing large quantities of starting substrate. Cardiac glycogen concentrations of diving turtles, for example, are nearly an order of magnitude higher than those observed in the hearts of terrestrial vertebrates, and this constitutes but one aspect of an impressive adaptation of this normally aerobic tissue for sustained anoxic function. By contrast, the flight muscle of the hummingbird contains large lipid droplets arranged regularly along an abundant battery of mitochondria, and this arrangement is but one aspect of the biochemical adaptation of this skeletal muscle for high aerobic rates of metabolism and work (see Drummond, 1971).

Since the above characteristics (kind and amount of substrate; kind and amount of enzyme) are variant, it is not surprising that the details of control circuitry also should be under stringent selection. Although the coarse level of control tends to be relatively conservative, the fine-control circuitry of muscle metabolism varies from muscle type to muscle type and is undoubtedly one of the most important design options available to organisms. Perhaps more than by any other mechanism, metabolism can evolve by adjustments in fine-control circuitry, almost always ultimately involving new interactions between key enzymes and metabolites either within the pathway or within some related pathway. When such new interactions arise they are either favoured or rejected by selection strictly on the basis of physiological utility. Hence, their significance can be fully appreciated only when the overall metabolic organization has been clarified and once the selective forces acting on it are known.

Since the environment is fundamentally unpredictable, built into the metabolic machinery must be a flexibility allowing for adjustments with respect to such factors as temperature and pressure, a flexibility that, as we have seen, often expresses itself in terms of constraints on the functional design of the system. Similar flexibility must be included with respect to other environmental factors which may impose directly on cell chemistry, such as the general salinity, CO_2, HCO_3^-, pH, other ions, metabolites and O_2 availability. The blend of anaerobic and aerobic metabolism in any given muscle is presumably an outcome of adjustments with respect to O_2 availability, but it is treated as a separate category (Table 1) because adaptation to the O_2 environment can lead to such profound metabolic reorganization that it in effect determines which other alternatives theoretically available to the organism are in fact utilized. For example, the type of substrate utilized often relates to the blend of anaerobic and aerobic metabolic pathways. A high anaerobic/aerobic ratio would tend to delete fat as a potential substrate source, probably emphasizing carbohydrate and/or amino acids, whereas by contrast, in the vertebrates at least, a high aerobic/anaerobic ratio is usually sustained by fat catabolism. As we have seen, the only known obligatorily aerobic muscles (squid mantle and insect flight muscles) seem to utilize two substrates (carbohydrate and proline) simultaneously. Be that as it may, although some sort of mix of anaerobic and aerobic metabolism must be utilized, it is clear that organisms have a wide spectrum of possibilities to work with and that the actual ratio utilized by any given species will depend on the availability of O_2 at the muscle cell during periods of work and rest. If problems are encountered, caused by a temporary lack of O_2 (as in burst-work fibres), organisms may utilize compensatory strategies of adaptation. Other organisms, by maximizing energetic efficiency and minimizing power output, are capable of exploiting anoxic environments on a sustained basis. A similarly exploitative strategy of adaptation is represented by organisms that design obligatorily aerobic metabolism to exploit fully the advantages of either an unlimited environmental supply or an unlimitable O_2-delivery system to the muscle or both.

Often, the choices made by organisms are intimately associated with the length of time available for the adaptation. This interplay between time and adaptive strategy is of two types. First, as we pointed out, time in the sense of an organism's phylogenetic history often determines which of several theoretically available options are in fact true choices. For example, the octopus mantle utilizes octopine dehydrogenase and arginine phosphate rather than lactate dehydrogenase and creatine phosphate, presumably because its phylogeny 'locked' it into these alternatives. Secondly, the time-course of environmental change strongly limits mechanisms of adaptation that an organism can harness. An organism's instantaneous response to limited O_2 (a compensatory activation of anaerobic metabolism) may be different from an acclimatization response (in which a larger fraction of the organism's lifecycle is involved and different mechanisms, such as adjustments in steady-state levels of enzymes, can be harnessed). Finally, phylogenetic time-periods (many generations of species' time) are required for the most exploitative of adaptational strategies, those based on a gradual reorganization of metabolism and on the redesign of catalytic and regulatory properties of enzymes for optimal function within both the internal and the external environment.

A biomechanic, Dr. Steve Wainwright, talked me into writing this essay. Many of the ideas expressed have been frequently discussed with my students; much of the experimental basis for the paper comes from their own studies.

References

Alexander, R. McN. (1967) *Functional Design in Fishes*, pp. 160. Hutchinson University Library, London
Atkinson, D. E. (1977) *Energy Metabolism and its Regulation*, Academic Press, New York, in the press
Collicutt, J. (1975) M.Sc. Thesis, University of British Columbia
De Zoeten, L. W. & Tipker, J. (1969) *Hoppe-Seyler's Z. Physiol. Chem.* **350**, 691–695
de Zwann, A., Kluytmans, J. H. F. M. & Zandee, D. I. (1976) *Biochem. Soc. Symp.* **41**, 133–168
Drummond, G. I. (1971) *Am. Zool.* **11**, 83–97
Fields, J. H. A. (1976) *Fed. Proc. Fed. Am. Soc. Exp. Biol.* **35**, in the press
Fields, J. H. A., Baldwin, J. & Hochachka, P. W. (1976) *Can. J. Zool.* in the press
Froede, H. C. & Wilson, I. B. (1972) *Enzymes* **5**, 89–114
Guderley, H. E., Storey, K. B., Fields, J. H. A. & Hochachka, P. W. (1976) *Can. J. Zool.* in the press
Hochachka, P. W. (1974) *Biochem. J.* **143**, 535–539
Hochachka, P. W. (1975) *Comp. Biochem. Physiol. B* **50**, in the press
Hochachka, P. W. & Storey, K. B. (1975) *Science* **187**, 613–621
Hochachka, P. W., Fields, J. H. A. & Mustafa, T. (1973) *Am. Zool.* **13**, 543–555
Hochachka, P. W., Owen, T. G., Allen, J. F. & Whittow, G. C. (1975a) *Comp. Biochem. Physiol. B* **50**, 17–22
Hochachka, P. W., Baldwin, J. & Storey, K. B. (1975b) *Comp. Biochem. Physiol. B* **50**, in the press
Hochachka, P. W., Baldwin, J. & Storey, K. B. (1975c) *Comp. Biochem. Physiol. B* **50**, in the press
Hochachka, P. W., Moon, T. W., Mustafa' T. & Storey, K. B. (1975d) *Comp. Biochem. Physiol. B* **50**, in the press
Hochachka, P. W., Norberg, C., Baldwin, J. & Fields, J. H. A. (1976) *Nature (London)* in the press
Holbrook, J. J., Liljas, A., Steindel, S. J. & Rossmann, M. G. (1975) *Enzymes* in the press
Luisi, P. L., Baici, A., Olomucki, A. & Doublet, M. O. (1975) *Eur. J Biochem.* **50**, 511–516
Mansour, T. E. (1972) *Curr. Top. Cell. Regul.* **5**, 1–46
Matsuura, D. T. & Whittow, G. C. (1973) *Am. J. Physiol.* **225**, 717–715
Moon, T. W. & Hulbert, W. C. (1975) *Comp. Biochem. Physiol. B* **50**, in the press
Mustafa, T. (1975) *Proc. Can. Soc. Zool.* 15–18
Owen, T. G. & Hochachka, P. W. (1974) *Biochem. J.* **143**, 541–553
Sacktor, B. (1970) *Adv. Insect Physiol.* **7**, 267–347
Sacktor, B. (1976) *Biochem. Soc. Symp.* **41**, 111–131
Safer, B. & Williamson, J. R. (1973) *J. Biol. Chem.* **248**, 2570–2579
Saz, H. J. (1971) *Am. Zool.* **11**, 125–135
Scrutton, M. C. & Utter, M. F. (1968) *Annu. Rev. Biochem.* **37**, 249–302
Simon, L. M., Robin, E. D., Elsner, R., van Kessel, A. L. G. J. & Theodore, J. (1974) *Comp. Biochem. Physiol. B* **47**, 209–215
Somero, G. N. & Low, P. S. (1976) *Biochem. Soc. Symp.* **41**, 33–42
Stevens, E. D. & Black, E. C. (1966) *J. Fish. Res. Bd. Can.* **23**, 471–485
Stokes, T. & Awapara, J. (1968) *Comp. Biochem. Physiol. B* **25**, 883–892
Storey, K. B. & Hochachka, P. W. (1974a) *J. Biol. Chem.* **249**, 1423–1427
Storey, K. B. & Hochachka, P. W. (1974b) *Comp. Biochem. Physiol. B* **49**, 119–128
Storey, K. B. & Hochachka, P. W. (1975a) *Comp. Biochem. Physiol. B* **50**, in the press
Storey, K. B. & Hochachka, P. W. (1975b) *Comp. Biochem. Physiol. B* **50**, in the press
Storey, K. B. & Hochachka, P. W. (1975c) *Comp. Biochem. Physiol. B* **50**, in the press

Biochem. Soc. Symp. (1976) **41**, 33–42
Printed in Great Britain

Temperature: A 'Shaping Force' in Protein Evolution

By GEORGE N. SOMERO and PHILIP S. LOW

Scripps Institution of Oceanography, University of California, San Diego,
Box 1529, *La Jolla, CA* 92093, *U.S.A.*

Synopsis

1. Comparisons of homologous enzymes from species adapted to widely different temperatures reveal that ligand-binding affinities are rigorously conserved. This is interpreted to mean that a critical relationship between ligand-binding ability and intracellular ligand concentrations must be maintained for proper enzymic regulation. 2. The catalytic efficiencies of enzyme homologues differ in temperature-compensatory manners. Activation free energies are proportional to adaptation temperature and, consequently, low-temperature-adapted enzymes have the highest substrate turnover numbers. 3. Temperature compensatory adjustments in catalytic efficiency may be achieved by altering the number of weak bonds that form or break during a catalytic conformational change. Support for this hypothesis comes from the finding that activation enthalpy and activation entropy values co-vary in a regular manner and by magnitudes consistent with different amounts of weak-bond formation/rupture during catalytic activation in different enzyme homologues. 4. Adaptive adjustments in ligand-binding energetics may also involve utilization of the energy changes that occur during conformational changes. This mechanism would permit enzymes with identical binding-site chemistries to display adaptively different ligand affinities. 5. The greater heat-stabilities of enzymes from warm-adapted species may cause these enzymes to be less efficient catalysts than cold-adapted heat-labile enzymes. Heat-stable enzymes may have to break more weak bonds during a catalytic conformational change than do cold-adapted enzymes. The requirements for thermal stability and high catalytic efficiency thus appear to force an adaptational 'compromise'.

Introduction

It is clear that most types of enzymes exist in tissue-specific isoenzyme forms. The kinetic properties of each specific isoenzyme generally reflect adaptation to the particular catalytic and regulatory roles the isoenzyme plays in the economy of the cell type in which it occurs.

In view of the large-scale effects of external environmental factors such as temperature and hydrostatic pressure on enzyme structure and function, one might expect to find environmentally specific variants of enzymes in organisms adapted to different environmental conditions, much as one finds tissue-specific enzyme forms within a single individual. The final 'design' of an enzyme therefore will reflect adaptation to both the internal and external environmental factors which affect its function and structure.

Whereas a great deal is known about the manners in which the metabolic functions of an enzyme 'shape' its properties, we are only now beginning to appreciate the extent to which external environmental factors influence enzyme evolution. Paramount among these latter factors is temperature, for in essence, temperature affects everything an enzyme is and does. Changes in temperature greatly alter rates of catalysis, and, perhaps more importantly, temperature changes can affect the weak bonds or secondary interactions that stabilize enzyme–ligand (substrate, modulator and cofactor) complexes and the higher orders of protein structure. One can ask, therefore, what similarities and differences exist in the catalytic, regulatory and structural properties of enzyme homologues from organisms adapted to different absolute temperatures and to widely different ranges of temperatures. What enzymic traits are rigorously conserved in all cases, and what traits are modified in temperature-compensatory manners? What are the focal points of enzymic adaptation to temperature over long, evolutionary time-spans?

We shall approach these and related questions in two steps. First, we shall discuss the enzymic kinetic properties which appear to be critically controlled during temperature-adaptation processes. Second, we shall examine possible molecular mechanisms which effect these kinetic adaptations. By elucidating the enzymic traits that are rigorously conserved under all environmental conditions, we will more fully understand the structural and functional properties of enzymes which are critically important for enzyme function. If we are able to discover how different enzyme homologues modify the same vital process, e.g., catalytic-rate enhancement, in temperature-compensatory manners, we may even be able to determine the fundamental mechanisms by which such an important process is achieved.

Focal Points in Enzymic Adaptation to Temperature

(1) *Conservation of ligand-binding affinities*

Proper affinities for substrates, cofactors and regulatory modulators are absolutely essential for optimal enzymic function, for an enzyme must not only be able to catalyse a reaction at a high rate, but it must also be able to vary its rate of catalysis according to the needs of the cell. The latter are reflected in the concentrations of substrates, modulators and cofactors present near the enzyme's surface, and the correct ability to bind the proper ligands is tantamount to an enzyme's being able to 'read' properly the metabolic state of its local environment.

Because enzyme–ligand complexes are stabilized by weak interactions such as hydrogen bonds, electrostatic interactions and hydrophobic interactions which form/rupture with small changes in energy, even slight perturbations of the cellular environment may exert large changes in the ligand-binding abilities of enzymes. For example, as temperature rises from approximately 5° to 20°C, the affinity of *Trematomus* pyruvate kinase for phosphoenolpyruvate decreases severalfold, as indicated by a rise in the apparent Michaelis constant (K_m) of phosphoenolpyruvate (Fig. 1). Since *Trematomus* lives at nearly constant temperatures below 0°C, this effect poses no problem. However, many ectotherms (poikilotherms) experience cell temperature ranges of 10–20°C, and for these species large temperature-induced changes in substrate-binding ability seem inimical to optimal regulatory function (Fig. 2). Too low a substrate-binding ability means that the enzyme would not utilize a significant portion of its catalytic potential, and the effect of a negative modulator on the enzyme would be a too rapid decrease in catalytic rate. Conversely, if the affinity for substrate is too high, the enzyme may

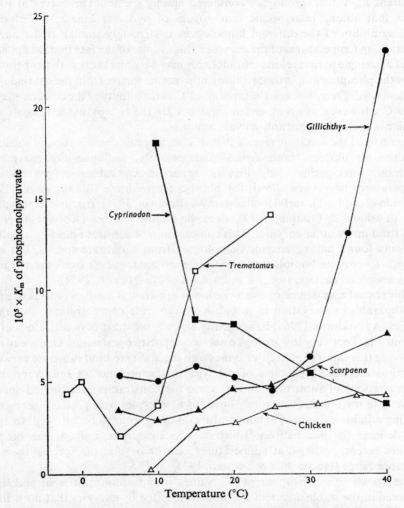

Fig. 1. *Effect of assay temperature on the apparent Michaelis constant (K_m) of phosphoenolpyruvate for skeletal-muscle pyruvate kinases (ATP–pyruvate phosphotransferase: EC 2.7.1.40) from different species*

Trematomus (Pagothenia) borchgrevinki (□) is an Antarctic teleost fish which lives at −1.86°C throughout the year. *Cyprinodon macularius* (■) is a warm-adapted eurythermal fish found in shallow streams and ponds in the deserts of the South-Western United States (Somero, 1975). The fish commonly occurs in waters with temperatures near, or just above, 40°C. *Gillichthys mirabilis* (●) is a eurythermal estuarine fish found along the South-Western coast of North America from approximately the latitude of San Francisco to Baja, California. The annual range of temperatures in its habitat is approximately 10–30°C. *Scorpaena gutatta* (▲) is a marine teleost which experiences temperatures of 7–18°C.

be rapidly saturated with substrate with the result that substrate concentrations may begin to rise uncontrollably in the cell.

In view of the inherent dangers in allowing substrate (and modulator and cofactor) affinities to vary widely from values which are optimal for regulation, it is hardly surprising to discover that substrate-binding abilities, as measured by apparent K_m values, are highly conserved among species. The curves of Fig. 1 show that among interspecific homologues of pyruvate kinase, the absolute binding ability of the different homologues is strikingly similar at the normal functioning temperatures of the enzymes, this in spite of the fact that the apparent K_m of any single pyruvate kinase homologue may vary markedly with temperature. Thus the phosphoenolpyruvate affinity of pyruvate kinase from the cold-adapted Antarctic fish *Trematomus* is the same at $-2°C$ as the affinity of the chicken enzyme at $39°C$. However, at no other temperatures are the two pyruvate kinase forms similar in their phosphoenolpyruvate affinities.

A survey of the literature reveals that this interspecific conservation of substrate affinities for different homologues (and, possibly, analogues) of enzymes is common. Interspecific similarities in apparent K_m values at physiological temperatures have been shown for phosphofructokinase (Blangy et al., 1968; Yoshida et al., 1971), acetylcholinesterases (Baldwin, 1971), fructose diphosphatases (Yoshida & Oshima, 1971), deoxythymidine kinases (Kobayashi et al., 1974) and many other enzymes. This conservation of substrate-binding affinity is not only found among enzyme homologues from vertebrate species, but even extends to enzyme homologues (possibly even analogues?) from thermophilic and mesophilic bacteria (see, for example, Kobayashi et al., 1974).

This remarkable degree of conservation in apparent K_m values reflects, we feel, a comparable conservation in intracellular substrate concentrations among all species. As Atkinson (1969) has cogently argued, substrate concentrations in the cell must be kept very low owing to solvent-capacity constraints. One means for achieving this end is to 'design' enzymes with the ability to bind substrate very well and thus facilitate high rates of catalysis in the presence of only very small concentrations of substrates. In this view, the constraints of a limited solvent capacity in the cellular water may have led to enzymes with precisely set ligand-binding affinities. Changes in temperature do not appear to be 'allowed' to upset this delicate balance between substrate concentrations and enzyme binding abilities except, perhaps, at temperatures near or beyond the species' thermal-tolerance range (Hochachka & Somero, 1973).

The conservation in apparent K_m values found among different species is mirrored in the stable apparent K_m values exhibited by enzymes that must function over wide temperature ranges in a single organism (Fig. 1). The K_m of phosphoenolpyruvate of the pyruvate kinase of *Gillichthys mirabilis*, a highly eurythermal estuarine fish, is essentially constant over the range of temperatures the species experiences (approximately 10–30°C). In contrast, the K_m of phosphoenolpyruvate of the enzyme from the stenothermal Antarctic fish *Trematomus* changes relatively fast as the temperature varies. Clearly, much as selection has 'shaped' all pyruvate kinases to have similar phosphoenolpyruvate affinities at physiological temperatures, enzymes that work over wide temperature ranges have been 'designed' to maintain this optimal affinity in a temperature-independent manner.

This latter achievement represents quite an evolutionary accomplishment on thermodynamic grounds. Phosphoenolpyruvate, like most other metabolic intermediates, is a charged and highly polar molecule which is held to the substrate-

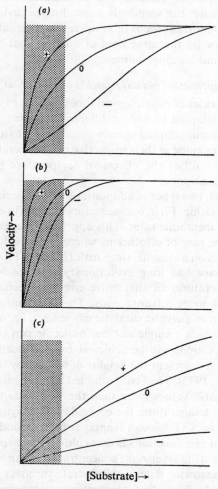

Fig. 2. *A diagrammatic illustration of the importance of preserving the correct substrate-binding affinity to ensure correct metabolic regulation*

Correct metabolic regulation includes the ability of the enzyme to respond to changes in substrate concentration by varying its rate of function over a wide range and the proper change in catalytic rate in response to the binding of a positive (+) or a negative (−) modulator. 0 indicates no bound modulator. (*a*) An enzyme having an optimal affinity for substrate (apparent K_m value). The enzyme can vary its activity over a wide range as substrate and modulator concentrations change. (*b*) An enzyme with too high an affinity for substrate. Here there exists a risk of saturating the enzyme when substrate concentrations rise or when a positive modulator binds to the enzyme. Once the enzyme is saturated with substrate, substrate concentrations will rise uncontrollably. (*c*) An enzyme with too low an affinity for substrate. Here the enzyme responds sluggishly to changes in substrate concentration, and high concentrations of substrate must be reached before the enzyme catalyses a high rate of product formation. Response to a negative modulator may be extreme, i.e. a signal to retard metabolism may lead to a virtual cessation of catalysis. Physiological substrate concentrations are indicated by stipling.

binding site via electrostatic interactions and hydrogen bonds. These weak interactions form exothermally and thus would be weakened by increases in temperature; yet we find no evidence for weakening of the phosphoenolpyruvate–pyruvate kinase complex of *Gillichthys* as temperature rises from 10° to 30°C. We will attempt to resolve this seemingly anomalous behaviour after we have first examined a second important focus of temperature adaptation, catalytic activation, which may draw on the same pool of adaptational 'raw material' that is utilized to adjust ligand-binding abilities.

(2) *Temperature compensation via adjustments of activation free energies*

If we compare the rates of physiological function, e.g. oxygen consumption, of ectothermic species adapted to widely different temperatures, it is common to find that cold- and warm-adapted species metabolize at strikingly similar rates, when comparisons are made at their respective cell temperatures (Bullock, 1955). These adaptations to offset the effects of temperature on rates are termed 'temperature compensations'.

At the enzyme level, two types of adaptations seem especially likely to facilitate temperature compensation. First, compensatory changes in enzyme concentration could readily adjust metabolic rates. This approach to metabolic compensation seems common in the case of ectotherms where an individual must acclimate to different temperatures on a seasonal time-scale (Hazel & Prosser, 1974). However, for species which have had long evolutionary time-spans to adjust to a new environmental temperature, an alternative enzymic mechanism of temperature compensation seems more advantageous. This mechanism involves the compensatory adjustment of enzymic catalytic efficiencies, i.e. it entails the modification of the rate at which a single enzyme molecule can convert substrate into product. These adaptations will be achieved by modification of activation free energies (ΔG^{+}), which represent the heights of the energy 'barriers' to metabolic reactions (Low et al., 1973). Thus cold-adapted homologues of an enzyme would have lowered their ΔG^{+} values more than the warm-adapted homologues and, consequently, at any temperature, the cold-adapted enzyme would facilitate the higher rate of catalysis. An obvious advantage of this second strategy for temperature compensation of rate functions is that cold-adapted species are not required to synthesize and retain higher enzyme concentrations than warm-adapted species.

The activation parameter data and substrate-turnover number ($V_{max.}$) data presented in Table 1 reveal that temperature compensatory adjustments in catalytic efficiency appear to be common among enzymes. In all cases examined to date, enzymes from low-cell-temperature species display lower ΔG^{+} values and higher substrate-turnover numbers than the homologous enzymes from warm-adapted species. The degree of temperature compensation that results from these changes in catalytic efficiency is indicated by the $V_{max.}$ data. For example, at 5°C, a tuna or halibut lactate dehydrogenase molecule can catalyse the conversion of pyruvate into lactate 3–4 times as rapidly on a per enzyme basis as rabbit lactate dehydrogenase.

It is important to realize that this temperature compensation of metabolic rates will pertain at all substrate concentrations, and not only under conditions of saturating ($V_{max.}$) substrate. Since interspecific homologues of enzymes have

Table 1. *Catalytic activation parameters (activation free energies, ΔG^+, enthalpies, ΔH^+, and entropies, ΔS^+) for homologous enzymic reactions of species adapted to different temperatures*

The data for pyruvate kinases were obtained by Low & Somero (unpublished work) and the data for the other three enzymes were taken from Low & Somero (1974). $V_{max.}$ is expressed as μmol of substrate converted into product/min per mg of enzyme at an assay temperature of 25°C. Except for the pyruvate kinase of the Antarctic fish *Trematomus borchgrevinki* the enzyme existed in two temperature-dependent conformational states. The transition temperatures ranged between 20° and 26°C, depending on the species; 5°C values are for the low-temperature conformation of pyruvate kinase; 30°C values are for the high-temperature conformer.

Pyruvate kinases	Cell temperature range (°C)	$V_{max.}$	ΔH^+ (cal/mol)	ΔS^+ (entropy units)	ΔG^+ (cal/mol)
Chicken	39	205			
5°C			17450	13.4	13720
30°C			11100	−8.0	13520
Rabbit	37	200			
5°C			15100	5.3	13640
30°C			11500	−6.7	13530
Bufo marinus (toad)	25–32	225			
5°C			14450	3.5	13490
30°C			11350	−7.0	13460
Mugil cephalus (mullet fish)	18–30	245			
5°C			13200	−0.5	13330
30°C			10950	−8.1	13410
Scorpaena gutatta (fish)	8–17	305			
5°C			13150	−0.6	13320
30°C			11150	−7.0	13280
Trematomus borchgrevinki (Antarctic fish)	−2				
5°C			11700		
30°C			11700		

	Assay temperature (°C)	$V_{max.}$	ΔH^+ (cal/mol)	ΔS^+ (entropy units)	ΔG^+ (cal/mol)
M₄ lactate dehydrogenases					
Rabbit	5	95	12525	−2.5	13230
	35	958	12525	−2.5	13310
Chicken	5	168	10500	−8.7	12920
	35	1184	10500	−8.7	13180
Tuna	5	355	8775	−13.4	12500
	35	1846	8775	−13.4	12900
Halibut	5	355	8770	−13.7	12500
	35	1826	8700	−13.7	12910
D-Glyceraldehyde 3-phosphate dehydrogenases					
Rabbit	5	6.1	15300	11.4	18450
	35	180	14900	11.3	18400
Lobster	5	22.7	14550	−2.2	13950
	35	220	14800	−2.9	13900
Cod	5	18.5	14700	−2.6	13950
	35	225	14800	−2.9	13900
Muscle glycogen phosphorylase *b*					
Rabbit	0	0.8	15950	17.2	20650
	30	60	15200	17.8	20600
Lobster	0	4.5	15050	1.1	15350
	30	70.8	15100	0.8	15300

similar substrate-binding abilities at physiological temperatures, the catalytic advantage displayed by a cold-adapted enzyme at $V_{max.}$ substrate concentrations will also be present at low physiological substrate concentrations.

The finding that differences in catalytic efficiency correlate with adaptation temperature and not with phylogenetic status in the case of pyruvate kinase homologues reveals that a significant degree of enzyme evolution related to temperature adaptation has occurred subsequent to the divergence of the major vertebrate lines (P. S. Low & G. N. Somero, unpublished work). Thus the colonization of different thermal habitats by fishes and amphibians has been marked by changes in enzymic catalytic efficiencies of the same order of magnitude as the changes that occurred during the evolution of high-cell-temperature enzymes in birds and mammals.

The Molecular Basis of Adaptations in Kinetic Properties

Let us now examine the possible molecular mechanisms that might be responsible for the observed adjustments in ligand-binding abilities and in activation parameters. For the latter case, a clue to the nature of the adaptational mechanism is provided by the manner in which the activation enthalpies (ΔH^+) and entropies (ΔS^+) co-vary among different homologues of each type of enzyme examined (Fig. 3). In the case of pyruvate kinase, for example, a change in ΔH^+ of 302 calories/mol is associated with a change in ΔS^+ of 1 entropy unit. A similar co-variation in ΔH^+ and ΔS^+ is observed for all other enzymes compared in this manner (Fig. 3). These results suggest the following important conclusion: the

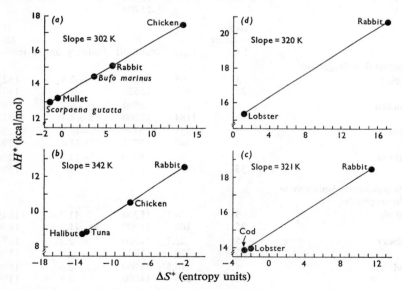

Fig. 3. *Patterns of co-variation of activation enthalpy (ΔH^+) and activation entropy (ΔS^+) for different homologues of pyruvate kinase (a), M_4 lactate dehydrogenase (b), glyceraldehyde 3-phosphate dehydrogenase (c) and glycogen phosphorylase b (d)*

The data for pyruvate kinase were obtained by Low & Somero (unpublished work). All other data are from Low & Somero (1974).

observed differences in activation parameters among enzyme homologues result from the occurrence of a single type of process which occurs to different extents in the different homologues. This process is characterized by a $\Delta H/\Delta S$ ratio of between 300 and 340 K (Fig. 3). In other words, the decreases in ΔG^+ values of cold-adapted enzymes have been accomplished by lowering ΔH^+ and ΔS^+ in a ratio near 320 K.

The types of molecular processes characterized by this $\Delta H/\Delta S$ ratio seem promising candidates for adjusting activation energetics. These processes are weak-bond formations (i) between protein groups (amino acid side chains and peptide linkages) within the enzyme (Low & Somero, 1974), and (ii) between protein groups and the surrounding solvent (Low & Somero, 1975a,b). These processes occur with large changes in free energy, enthalpy and entropy, which would contribute significantly to the activation characteristics of a reaction, provided that the weak bonds were formed or broken during the activation event. Such changes in weak interactions are likely during catalytic conformational changes (Koshland & Neet, 1968; Low & Somero, 1975b).

The adjustment of ligand-binding energetics may involve similar mechanisms (Somero & Low, 1976). Since ligand binding is commonly associated with conformational changes in the enzyme (Koshland & Neet, 1968), the energy changes that result from weak-bond formation/rupture between amino acid residues within the enzyme or between amino acid residues and the surrounding water may 'titrate' the energy changes that occur locally at the ligand-binding site. For example, if substrate binding is strongly exothermic, the pairing of this reaction with an endothermic conformational change, e.g. one which involves the rupture of intraprotein hydrogen bonds, could decrease the enthalpy of binding to a value near zero. An important aspect of this mechanism is that enzymes with identical substrate-binding-site chemistry may yet have adaptively different ligand-binding characteristics (Somero & Low, 1976).

In conclusion, the conformational changes which accompany catalytic activation and ligand binding may be extremely important in 'shaping' the energy changes which accompany these processes. Temperature adaptation may involve the 'fine tuning' of these conformational changes to adjust the accompanying energy changes to values which are optimal on biological grounds.

Enzyme Structural Stability and Temperature Adaptation

In addition to being a sensitive regulatory agent and an efficient catalyst, an enzyme must be able to maintain its native structure under the temperature regime it experiences. Indeed, temperature adaptation appears to have led to alterations in enzyme thermal-stability in many cases. For example, the heat-denaturation temperature of an enzyme is often correlated with adaptation temperature (see, e.g., Smith, 1973a,b). Pyruvate kinases also display this relationship (P. S. Low & G. N. Somero, unpublished work).

In view of the fact that adaptation to high temperatures involves the gaining of heat-stability and the loss of some catalytic efficiency by enzymes, one can ask whether this correlation is due to two independent paths of enzyme evolution, or whether some common structural factor links both changes. It is tempting to speculate that structural strength, as estimated by heat-denaturation temperatures and catalytic efficiency are closely related, i.e. are partly determined by a common factor. If the energy changes that occur during a catalytic conformational change contribute significantly to the activation free energy, as we have suggested above, then an enzyme that is stabilized by relatively large numbers of secondary interactions may have to 'work harder' to effect a catalytic conformational change than does an enzyme with a more flexible structure. Thus if an enzyme from a warm-adapted organism has to break, on the average, one additional hydrogen bond during catalysis relative to an enzyme from a cold-adapted species, we would expect that the activation parameters (ΔH^+, ΔS^+, ΔG^+) would all be higher for the reaction catalysed by the warm-adapted enzyme.

This loss in catalytic efficiency experienced by a warm-adapted enzyme must be viewed in a sufficiently broad context, however. The gain in heat-stability by the warm-adapted enzyme may be of greater benefit than the retention of a higher catalytic efficiency. Since rates of catalysis will be accelerated by temperature increases, the warm-adapting enzyme will not suffer significant losses in catalytic rates. For example, a rabbit lactate dehydrogenase molecule works approximately three times as fast at 35°C as a tuna or halibut lactate dehydrogenase molecule works at 5°C. The more rigid structure of the warm-adapted enzyme may also help to maintain the integrity of the ligand-binding sites' geometries at high temperatures.

In conclusion, the 'shaping' of enzymic properties during adaptation to different thermal regimes involves the modification of a wide number of structural and functional characteristics of the enzyme, and these adaptations often appear to reflect the striking of important adaptational 'compromises'.

These studies were supported by National Science Foundation grants GB-31106 and BMS 74-17335.

References

Atkinson, D. E. (1969) *Curr. Top. Cell. Regul.* **1**, 29–43

Baldwin, J. (1971) *Comp. Biochem. Physiol. B* **40**, 181–187

Blangy, D., Buc, H. & Monod, J. (1968) *J. Mol. Biol.* **31**, 13–35

Bullock, T. H. (1955) *Biol. Rev. Cambridge Philos. Soc.* **30**, 311–342

Hazel, J. & Prosser, C. L. (1974) *Physiol. Rev.* **54**, 620–677

Hochachka, P. W. & Somero, G. N. (1973) *Strategies of Biochemical Adaptation.* p. 358, W. B. Saunders and Co., Philadelphia

Kobayashi, S., Hubbell, H. & Orengo, A. (1974) *Biochemistry* **13**, 4537–4543

Koshland, D. E. & Neet, K. E. (1968) *Annu. Rev. Biochem.* **37**, 359–410

Low, P. S. & Somero, G. N. (1974) *Comp. Biochem. Physiol. B* **49**, 307–312

Low, P. S. & Somero, G. N. (1975a) *Proc. Natl. Acad. Sci. U.S.A.* **72**, 3014–3018

Low, P. S. & Somero, G. N. (1975b) *Proc. Natl. Acad. Sci. U.S.A.* **72**, 3305–3309

Low, P. S., Bada, J. L. & Somero, G. N. (1973) *Proc. Natl. Acad. Sci. U.S.A.* **70**, 430–432

Smith, C. L. (1973a) *Comp. Biochem. Physiol. B* **44**, 779–788

Smith, C. L. (1973b) *Compl Biochem. Physiol. B* **44**, 789–801

Somero, G. N. (1975) in *Environmental Physiology of Desert Organisms* (Hadley, N., ed.), Dowden, Hutchinson and Ross, Stroudsburg, PA

Somero, G. N. & Low, P. S. (1976) *Am. Nat.* in the press

Yoshida, M. & Oshima, T. (1971) *Biochem. Biophys. Res. Commun.* **45**, 495–500

Yoshida, M., Oshima, T. & Imahori, K. (1971) *Biochem. Biophys. Res. Commun.* **43**, 36–39

Biochem. Soc. Symp. (1976) **41**, 43–60
Printed in Great Britain

Temperature Adaptation in Fish

By M. W. SMITH

*Agricultural Research Council Institute of Animal Physiology, Babraham,
Cambridge CB2 4AT, U.K.*

Synopsis

Altering the body temperature of fish causes immediate changes in both the functional and biochemical properties of secretory epithelial tissues. The present account describes how the intestinal tract of the goldfish adapts to regulate its function after a sudden exposure to a change in environmental temperature. Adaptation can be seen to take place in three distinct steps. The first stage, occurring 15–20h after the change in environmental temperature, alters the ability of actively transported non-electrolytes to increase the microvillar membrane permeability to sodium. The passive permeability of the cells to sodium is also regulated at this stage. The second adaptive change involves changes in membrane phospholipids and amino acid transport. This occurs after about 32–48h. Regulation is still not complete, however, the intracellular sodium remaining low and intracellular potassium remaining high for 2–3 weeks. There is then a final regulation of sodium-pump activity. This change, which is concerned with pump turnover rather than with number of pump sites, is probably connected with the synthesis of new cells. These changes are discussed in detail, and an attempt is made to relate fine changes in phospholipid fatty acyl composition to known changes in membrane function.

Introduction

It is an accepted fact that the phenomenon that we call temperature adaptation involves changes both in the amounts and catalytic properties of different intracellular enzymes. It is still a matter of theory, however, as to exactly why these changes take place. In general it has been assumed that induced changes in enzyme properties are organized with the sole purpose of restoring the physiological state of the animal to that found immediately before the change in environmental temperature. If the metabolism or physiological function returns to that measured immediately before applying the external stimulus, this is then referred to as an ideal compensation. If they show some return, but still remain different from the original state, this is called a partial compensation (Precht, 1958). These terms are, to some extent, unfortunate in that they tend to impose false limits both on the purpose and efficiency of adaptation.

The primary purpose of adaptive change must be to survive. The coincidence or otherwise of survival with the complete return of an enzyme's activity to some arbitrarily defined normal level ought to be of secondary importance. Temperature adaptation of cellular metabolism does, nevertheless, usually result in some return towards a pre-existing pattern and I do not want to imply that this homoeostatic response is without importance. There must, however, be a place within this pattern for temperature-dependent differences to exist and these differences should not be looked on merely as something that has to be tolerated by the adapting organism.

The positively beneficial effects, which arise from maintaining differences in metabolic activity and physiological function, can best be appreciated by considering some of the long-term effects of temperature on the physiology of different animals. It is well known, for example, that the intake of food by fish is much increased as they adapt to a warm environment. Their rate of growth increases and they reach sexual maturity at an earlier age. Here we are dealing with a survival advantage in broader terms but the point I want to consider specifically is the role the intestine might play in such a situation. One might predict that some change in intestinal absorption and metabolism would be needed to sustain these temperature-dependent differences in growth rate. The rate of protein synthesis in the goldfish intestine does, in fact, increase by an order of magnitude for a 30°C increase in environmental temperature, all values being determined for fully adapted fish maintained at different fixed temperatures (Morris & Smith, 1967). There are also both quantitative and qualitative changes in the ability of the goldfish intestine to absorb different amino acids (Mepham & Smith, 1966; Smith, 1970). In this case temperature adaptation can be considered to act in two ways, first to ensure that the intestine will continue to function in an integrated manner at a high environmental temperature, and secondly to enable the intestine to take positive advantage of a generally improved situation for growth and development.

There is a further complication to studies of temperature adaptation, which is not normally appreciated. If we allow that the maintenance of overall homoeostasis can involve the establishment of different steady states as well as a return to the *status quo* and that the fully adapted state can involve major changes in cellular function, then it is quite possible to imagine that early adaptive responses might cause secondary, less important, changes in intracellular enzyme activities. Enzymes made redundant by an earlier adaptive change, say in the transport properties of a cell plasma membrane, might then cease to be synthesized. Conclusions about the biological significance of changes in enzyme activity, seen only from analysis of tissues taken from animals adapted to different temperatures over prolonged periods of time, could then be in error. It is much safer to follow the time-course of adaptive responses before coming to any conclusion about their physiological and biochemical significance. These procedures have been followed in studying the effect of temperature on transport processes in the goldfish gall bladder and intestine. I want to limit this review to a description of this type of experiment, concentrating mainly on the functional aspects of adaptation in these tissues. It is hoped, however, that concepts formed as a result of this type of work will find a wider application in describing some of the processes concerned with temperature adaptation in fish generally.

Studies on Goldfish

The gross structure of the goldfish intestine is much simpler than that of mammals, the mucosal layer constituting some 50% of the total wet weight. The intestine of the goldfish is also much more stable *in vitro* and in this respect it resembles other tissues taken from different poikilothermic animals. The ultrastructure of the goldfish intestinal mucosa is similar to that found in other species and so are the mechanisms whereby the intestine absorbs ions, water and different non-electrolytes.

A diagrammatic representation of an intestinal absorptive cell, together with an even simpler representation of how sodium and an actively transported sugar might cross the mucosa, is shown in Fig. 1. The cell is taken from the middle third of a jejunal villus of a normal man (Trier, 1968), and the model is derived from work carried out on the rabbit terminal ileum (Schultz & Zalusky, 1964), but the descriptions are equally valid for the fish intestine. The two diagrams are shown together to emphasize the simplifications used in describing this system. Material absorbed by the intestine has, in fact, nine potential barriers to cross

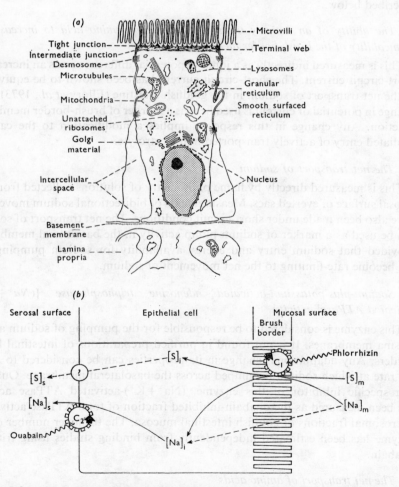

Fig. 1. *Schematic diagram of an intestinal absorptive cell (Trier, 1968) (a) and a model scheme proposed to account for the interactions that occur between the active transport of sodium and the active transport of sugars (Schultz & Zalusky, 1964) (b).*

before reaching the capillary blood supply (Palay & Karlin, 1959). Only two of these barriers are shown in the model. Sugars and amino acids are transported into the mucosal cell in association with sodium on carriers situated within the brush-border membrane (barrier 1). Sodium entering the cell, by diffusion or in association with an actively transported non-electrolyte, is pumped across the basolateral membrane (barrier 2) towards the serosal surface. Sugars and amino acids leaving the mucosa do so by a process of carrier-mediated facilitated diffusion. The model assumes that the kinetics of transport will not be influenced by the cell cytoplasm acting as a barrier to diffusion and that steps distal to the mucosal cell, e.g. crossing the basement membrane, the lamina propria and the endothelial cells of the intestinal capillaries, will not be rate-limiting to the overall process of absorption. Four different parameters of intestinal function, measured throughout a period of adaptation from a cold to a warm environment, are described below.

(1) *The ability of an actively transported sugar or amino acid to increase the permeability of the brush-border membrane to sodium*

This is measured indirectly as a jump in potential difference or as an increase in short-circuit current. The short-circuit current has been shown to be equivalent to the net transport of sodium in the goldfish intestine (Ellory *et al.*, 1973). The change in potential or current is used here as a marker of brush-border membrane function. Any change in this response implies a modification to the carrier-mediated entry of actively transported non-electrolyte.

(2) *The net transport of sodium*

This is measured directly by flame photometry of solutions collected from the serosal surface of everted sacs. Measurements of bidirectional sodium movement have also been made under short-circuit conditions. The net transport of sodium can be used as a marker of sodium-pump activity in the basolateral membranes provided that sodium entry and energy availability for sodium pumping has not become rate-limiting to the net movement of sodium.

(3) *Sodium-plus-potassium-activated   adenosine   triphosphatase   [$(Na^+ + K^+)$-activated ATPase] activity*

This enzyme is considered to be responsible for the pumping of sodium across plasma membranes. It is not found in purified preparations of intestinal brush borders. Any adaptational change in its properties can be considered to affect the rate at which sodium is pumped across the basolateral membrane. Ouabain is a specific inhibitor of this enzyme. $(Na^+ + K^+)$-activated ATPase activity has been measured as the ouabain-inhibited fraction of total ATPase activity in microsomal fractions of goldfish intestinal mucosa. The turnover number of the enzyme has been estimated independently from binding studies using tritiated ouabain.

(4) *The net transport of amino acids*

This is measured directly by Moore & Stein (1954) analysis of material collected from the serosal surface of preparations of goldfish intestine *in vitro*. Its use is to verify that adaptational changes in amino acid transport have taken place. Any adaptational change to the facilitated exit of amino acids will also be detected by this kind of measurement provided that the entry of amino acids can be shown not to become rate-limiting to net transport.

The fatty acid composition of phospholipids extracted from microsomal fractions of goldfish intestinal mucosa has also been determined, both for the fully adapted state and during the period when adaptation is taking place. This work has been carried out in collaboration with Dr. P. Kemp and Mr. N. Miller.

The goldfish gall bladder has also been used as a model system to study temperature adaptation in the goldfish. Its structure is similar to that of the intestine as is its main function, the reabsorption of sodium and water. There is, however, only occasional evidence for the active transport of non-electrolytes by this tissue and no evidence to suggest that the presence of non-electrolytes will support an electrogenic transport of sodium. The simplicity of this system allows one to dissect the adaptational response a little further. Measurements of net sodium and fluid transport show regulation with a time-course identical with that seen for the intestine (Cremaschi et al., 1973). Measurements of intracellular sodium and potassium concentrations in the gall-bladder epithelium during a period of adaptation then allows one to make further assumptions about the adaptational response in the intestine.

Fig. 2 shows how the effect of glucose, in increasing the electrogenic transport of sodium, changes when cold fish are adapted to a warm environment. The potential difference is dependent on the presence of glucose, increasing at incubation temperatures higher than the previous environmental temperature of the fish. The switching point for fish adapted to 15°C is 17.4°C for the example shown (Fig. 2b). This switching point increases to 30.8°C when the fish is fully

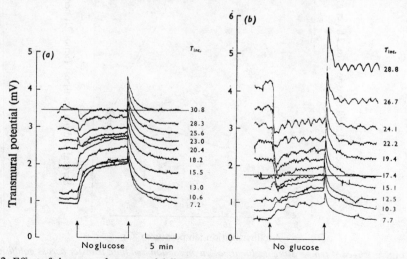

Fig. 2. *Effect of glucose on the potential difference maintained by goldfish intestine incubated in vitro at different temperatures*

The fish had been adapted previously to either 15°C (a) or 30°C (b). $T_{inc.}$ gives the incubation temperatures at which glucose effects were measured. The horizontal lines show temperatures above which glucose causes a permanent increase in potential (i.e. sodium movement).

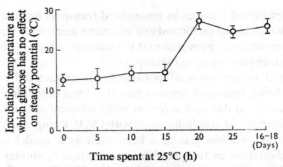

Fig. 3. *Temperature adaptation of a glucose response measured in goldfish intestine*

Glucose begins to stimulate the electrogenic transport of sodium, measured indirectly as a rise in potential difference across the tissue, at an incubation temperature which depends on the adaptational state of the fish. Goldfish adapted originally to 8°C were placed at 25°C and the time-course of changes was determined subsequently (Smith, 1966a).

adapted to 30°C (Fig. 2a). The time-course for this temperature-dependent change in membrane function is shown in Fig. 3. In this case the cold goldfish has been adapted to 8°C and the switch is to an environmental temperature of 25°C. The incubation temperature above which glucose stimulates sodium transport in the cold-adapted fish is about 12°C. This continues for up to 15h after the environmental temperature has been raised. Adaptation is complete some 5h later. The

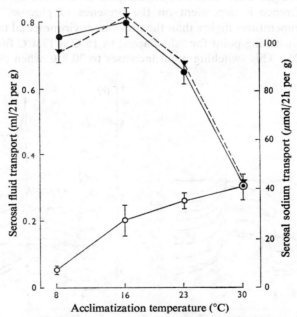

Fig. 4. *Effect of temperature acclimatization on sodium and fluid transfer across goldfish intestine*

Fish fully adapted to different environmental temperatures. ●, Transfer of fluid at 30°C; ○, transfer of fluid measured at incubation temperatures equal to the previous environmental temperature of the fish; ▼, transfer of sodium at 30°C. Mean values are given ±s.D. (Smith, 1970).

switching temperature is then about 25°C. This then remains constant for a period of at least 18 days. The prior injection of puromycin stops the subsequent change in membrane properties (Smith, 1966b).

The above results imply that the movement of sodium and possibly non-electrolytes is being regulated in the brush-border membrane as an early event in adaptation. Fig. 4 shows that the net transfer of sodium is much diminished, for constant incubation temperature at a high environmental temperature, and that the movement of water follows that of sodium. The medium transported to the serosal side of both intestine and gall bladder does, in fact, remain iso-osmotic with the mucosal medium at all stages of adaptation. The time-course with which this new steady state is achieved is shown in Fig. 5. In this case the cold fish had been adapted to 16°C and the environmental temperature changed to 30°C. The transfer of sodium and water measured 6–12h after changing the environ-mental temperature was not different from that obtained with intestines taken from fish adapted to 16°C. Adaptation to a new steady state took place some time within the next 12h, that is at about the time changes were occurring within the brush-border membrane.

The situation with regard to intracellular cation concentrations, measured dur-ing a similar period of adaptation, is shown for the goldfish gall bladder in Fig. 6. The regulation of fluid transport takes place with the same time-course as that seen in the intestine. The intracellular concentration of potassium is initially high (the intracellular concentration of sodium is correspondingly low) and this state of affairs persists for at least 2 days after raising the environmental temperature.

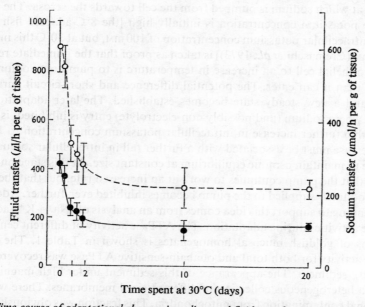

Fig. 5. *Time-course of adaptational changes in sodium and fluid transfer across goldfish intestine*

Fish, adapted originally at 16°C, were placed at 30°C for different periods of time. ○, Transfer of sodium; ●, transfer of fluid. Temperature of incubation was 30°C. Values are means±s.e. (Smith & Ellory, 1971).

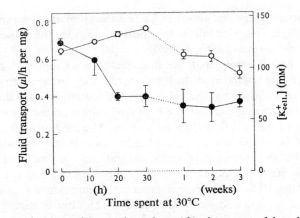

Fig. 6. *Time-course of adaptive changes taking place within the mucosa of the goldfish gall bladder*

Fish fully adapted to 15°C were placed at 30°C and their gall bladders removed at fixed times later. The transfer of fluid across these gall bladders was measured at 30°C and the intracellular concentration of potassium determined in mucosal scrapings. ●, Transfer of fluid; ○, intracellular concentration of potassium. Values are means±s.E. (Cremaschi *et al.*, 1973).

There is in addition some suspicion that the intracellular concentration of potassium is highest some 30 h after changing the temperature.

What can we conclude from this type of experiment? The initial act of raising the environmental temperature might be expected to do two things, to increase the diffusional and mediated entry of sodium into the mucosal cell and to stimulate the rate at which sodium is pumped from the cell towards the serosa. The finding that the potassium concentration is initially high [the 8°C-adapted fish at 8°C has an intracellular potassium concentration of 100 mM, but at 30°C this increases to 130 mM (Cremaschi *et al.*, 1973)] is taken as proof that the immediate response of the epithelial cell to an increase in temperature is to pump out sodium more quickly than it can enter. The potential difference and short-circuit current increase and a new steady state becomes established. The later adaptational response, where sodium (and possibly non-electrolyte) entry is inhibited, is associated with a further increase in intracellular potassium concentration to 140 mM. This increase must be associated with a further fall in intracellular sodium if the cells are to maintain osmotic equilibrium at constant size. At this stage one might predict that the pump continues to work at an increased rate, but that the rate at which sodium is supplied to the pump becomes inhibited even further. Additional work tending to support this idea comes from an analysis of $(Na^+ + K^+)$-activated ATPase activities during adaptation. The ATPase activity of different centrifugal fractions of goldfish mucosal homogenates is shown in Table 1. The highest specific activity for both total and ouabain-sensitive ATPase was recovered from a 20000 g sediment. The appearance of this sediment under high magnification was of a heterogeneous collection of different-sized membranes. There was only occasional contamination from mitochondria. This fraction was obviously a good source of basolateral membranes, but it was also contaminated to some extent by other types of membrane. The specific activity of the ouabain-sensitive ATPase decreased considerably when the environmental temperature was raised from 8°

Table 1. *ATPase activity of different centrifugal fractions of goldfish mucosal homogenates*

Values are from fish adapted to different environmental temperatures for a period of 2–3 weeks.
Ouabain concentration was 0.1 mM; incubation was at 37°C (Smith, 1967).

| Acclimatization temperature (°C) ... | ATPase activity (μmol of P_i/h per mg of protein) | | | | |
| | Total hydrolysis | | | Ouabain-sensitive hydrolysis | |
	8	19	30	8	19
Homogenate	20.4	21.1	21.3	4.38	4.05	4.23
10000g sediment	20.0	21.1	23.1	2.90	2.71	4.30
20000g sediment	36.8	34.5	42.2	20.1	12.3	11.4
85000g sediment	8.03	11.9	11.7	3.24	4.86	3.70
85000g supernatant	2.17	2.17	2.27	0.16	0.09	0.05
Overall recovery (as % of homogenate±S.E.M.)	59.0 ±2.5	56.1 ±1.6	63.2 ±6.6	81.0 ±21.0	58.1 ±6.3	62.4 ±8.6

to 30°C. The time-course leading to this change in specific activity is shown for a
separate series of experiments in Fig. 7.

The ouabain-sensitive ATPase activity of membrane fragments prepared from
goldfish intestinal mucosa showed no change during the first 2 days that the fish
were maintained at a high temperature. The sodium pump can be considered to be
capable of operating normally during this time-period, although this covers the
time when sodium transport is being regulated. Thus we have further evidence for
suggesting that it is the diffusional and carrier-mediated influx of sodium that is
first subject to adaptive control in this tissue. The ouabain-sensitive ATPase
activity, measured 10 days after raising the environmental temperature, is
somewhat less than in the cold-adapted animal. This difference becomes significant some time between 10 and 20 days after the initial change in environmental
temperature.

The ATP-dependent binding of tritiated ouabain to microsomal membranes
prepared from 16°C-adapted fish and from similar fish kept 20 days at 30°C was
not different [137±7 and 121±16pmol/mg of protein respectively (Smith &

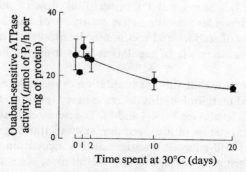

Fig. 7. *Ouabain-sensitive ATPase activities of goldfish intestinal membranes measured during
adaptation from 16°C to 30°C*

The concentration of ouabain used was 10μM. Each point is a mean±S.E. (Smith & Ellory,
1971).

Ellory, 1971)]. The turnover number of the enzyme decreased significantly (3505 to 2370 min^{-1}). This method for determining the number of enzyme molecules depends for its accuracy on the complete separation of non-specific from specific ouabain binding. With this reservation in mind we can conclude, tentatively, that the fall in ($Na^+ + K^+$)-activated ATPase activity does not result from a fall in the quantity of enzyme, but rather from a fall in the rate at which this enzyme operates. It is in any case a long time before this change becomes apparent. It could be associated with the appearance of a new cell population and it has nothing to do with the early regulation of sodium transport seen to take place in this tissue.

Before leaving this subject, however, it is important to mention that the intracellular concentrations of sodium and potassium, judged by the work on goldfish gall bladder, will not reach their final equilibrium values until the ($Na^+ + K^+$)-activated ATPase has changed its characteristics. Both intracellular potassium concentrations in goldfish gall bladder and ($Na^+ + K^+$)-activated ATPase activities in intestine take about 3 weeks to attain their final values. Protein synthesis is very susceptible to intracellular potassium concentration (Lubin, 1964) and one must assume that unusual things are happening to protein synthesis throughout this period of adaptation.

The preceding results allow one to speculate as to how ion transport and intracellular ion activities might be regulated in the intestine. I now want to describe some of the changes taking place in amino acid transport. We have seen that the immediate effect of an increase in environmental temperature is to increase the sodium gradient across the cell plasma membrane. It is generally accepted that this gradient is involved in providing the energy needed to move amino acids into the cell against their respective concentration gradients (Schultz & Curran, 1970). The regulation of sodium transport involves no major change in this gradient and one might suppose, if this were the only criterion for active transport, that adaptation of amino acid transport would not accompany regulation of sodium movement. Interactions between actively transported non-electrolytes and sodium taking place within the brush-border membrane are, nevertheless, responsive to adaptive control and it is probably this which causes the type of regulation seen for methionine transport in Fig. 8. Here the fall in net transfer is accompanied by a fall in net water transfer. Bulk flow of fluid is, under these conditions, itself dependent on the active transport of sodium. The connexion between adaptation of sodium and methionine transport can be explained, if it is accepted that the primary site of regulation lies within the brush-border membrane.

It is important to note that, though adaptation causes a decrease in the transfer of both sodium and methionine, this decrease never results in the rates returning to their pre-existing levels (see Figs. 4 and 8). The new steady states established at high environmental temperatures remain several times greater than those measured by using cold-adapted intestines at low incubation temperatures.

The final equilibrium concentration of methionine, measured at the serosal side of everted preparations of goldfish intestine, does not depend on the previous environmental temperature of the fish (Fig. 8). This may be because the adaptational response in the brush-border membrane affects mainly the binding of sodium and not amino acid to the ternary carrier. This idea has yet to be tested.

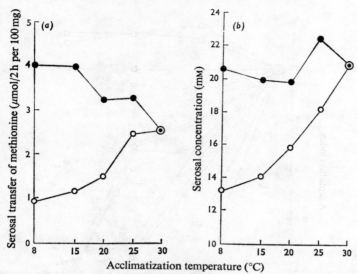

Fig. 8. *The effect of temperature adaptation on different aspects of methionine transfer across goldfish intestine*

Solutions containing 10mM-methionine were placed on both sides of everted sacs at the start of incubation and the serosal concentration was determined 2h later. ●, Serosal transfer (*a*) or serosal concentration (*b*) measured at 30°C; ○, serosal transfer (*a*) or serosal concentration (*b*) measured at incubation temperatures equal to the previous environmental temperatures of the fish. Each point gives the mean of six determinations (Smith, 1966*b*).

Some amino acids behave differently from methionine. There is, in fact, a whole spectrum of adaptive change in the power to transport amino acids out of the cell at concentrations that remain independent of the previous environmental temperature. Results illustrating this point have been summarized in Fig. 9. Values have been normalized to facilitate comparisons between different amino acids. The whole population of amino acids can be divided into one of two types. The first type (histidine, phenylalanine, methionine, tyrosine, leucine) is concentrated within the intestine and transported to the serosal solution at concentrations that are not greatly affected by changes in environmental temperature. The second type (proline, serine, valine, alanine, arginine, lysine, isoleucine, threonine) is also concentrated within the intestine, but the final concentration recovered from the serosal solution is markedly dependent on the previous environmental temperature of the fish. The first group of amino acids are known to be generally more lipophilic than the second and this has led to the notion that the lipid composition of the membranes might be able to exert a selective influence on amino acid diffusion. Adaptational changes are known to take place in the fatty acid composition of goldfish intestinal lipids. These are summarized in Table 2.

Analysis of two unidentified phospholipids (PY and PX) show negligible change in fatty acid composition. The major changes are confined to phosphatidylcholine and phosphatidylethanolamine. On raising the environmental temperature both these compounds show a pronounced increase in $C_{18:0}$ and a corresponding decrease in $C_{22:6}$ acyl chains. Phosphatidylserine and phosphatidylinositol, collected from t.l.c. plates as a single spot, show none of these changes.

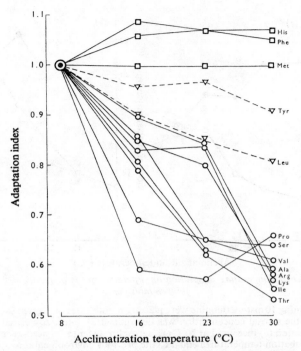

Fig. 9. *Effect of adaptation on the ability of goldfish intestine to concentrate amino acids at its serosal surface*

Each amino acid was used at an initial concentration of 0.5 mM. The final concentration reached by an 8°C-adapted intestine incubated at 30°C has been arbitrarily taken to be equal to 1. Other concentrations are then related to this reference. Broken lines for tyrosine and leucine indicate amino acids which partially resist this aspect of adaptation (Smith, 1970).

Table 2. *Analysis of different phospholipids extracted from membrane fractions prepared from goldfish adapted to different temperatures*

PX and PY, mixtures of unidentified phospholipids appearing on t.l.c. in regions corresponding to lysophosphatidylcholine and phosphatidic acid respectively; $T_{accl.}$, temperature of acclimatization (Kemp & Smith, 1970).

Phospholipid	$T_{accl.}$ (°C)	% of total phospholipid	Fatty acid composition (%)				
			$C_{18:0}$	$C_{20:1}$	$C_{20:3}$	$C_{20:4}$	$C_{22:6}$
PY	8	8.0	15.2		5.0	3.5	8.7
(band 1)	30	7.8	16.9		6.0	3.1	4.6
Phosphatidylethanolamine	8	26.7	10.9	4.2	3.7	13.9	23.3
(band 2)	30	27.3	27.0	1.1	5.2	12.2	12.8
Phosphatidylcholine	8	54.3	9.4	1.8	7.6	9.7	10.1
(band 3)	30	50.6	11.7	0.8	7.6	7.6	5.2
Phosphatidylserine + phosphatidylinositol	8	6.3	19.2		8.0	6.5	10.5
(band 4)	30	9.2	32.5		7.6	6.3	11.5
PX	8	4.7	16.7		8.4	3.7	4.1
(band 5)	30	5.0	17.7		7.5	1.4	2.8

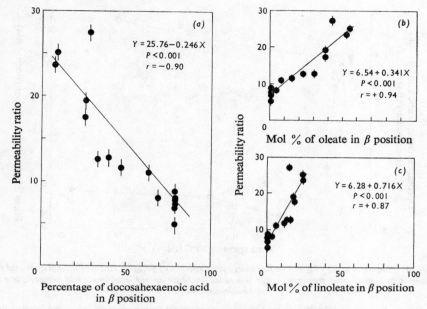

Fig. 10. *Relative permeability of liposomes to phenylalanine and glycine*

Liposomes were composed of phosphatidic acid, cholesterol and phosphatidylcholine. The fatty acid composition of the phosphatidylcholine was varied to give different proportions of docosahexaenoic acid in the β position. Relative permeability is plotted as the ratio of efflux constants, phenylalanine/glycine (Klein *et al.*, 1971).

In a separate series of experiments liposomes were made containing cholesterol, phosphatidic acid and phosphatidylcholine, the phosphatidylcholine being chosen to have different proportions of $C_{22:6}$ acyl chains, so as to correspond to membranes prepared from fish adapted to different environmental temperatures. Radioactively labelled amino acids were trapped within these liposomes at the time of formation and the rate constants for their subsequent efflux then determined in the normal way. Phenylalanine was chosen to represent group 1-type amino acids and glycine was chosen to represent group 2. The efflux rate for phenylalanine, which remained fairly independent of fatty acyl composition, was always considerably greater than that for glycine. The ratio of efflux rate constants phenylalanine/glycine is shown in Fig. 10. Decreasing the percentage of $C_{22:6}$ in the lipid bilayer caused a parallel increase in the phenylalanine/glycine efflux ratio, this being mainly due to a pronounced fall in the efflux rate constant for glycine. The main acyl group replacing $C_{22:6}$ was $C_{18:1}$ and one can derive another correlation with the efflux ratio phenylalanine/glycine being highest for highest concentrations of $C_{18:1}$. This way of presenting results might be important in view of the known ability of $C_{18:1}$ to form condensed lipid films with cholesterol (Ghosh *et al.*, 1971). The conclusion to be drawn from work with this purely artificial system is that changes in the phospholipid fatty acid composition can cause selective changes in the way amino acids diffuse across membranes. These changes are also in the same direction as would be predicted from previous work with the goldfish intestine.

3

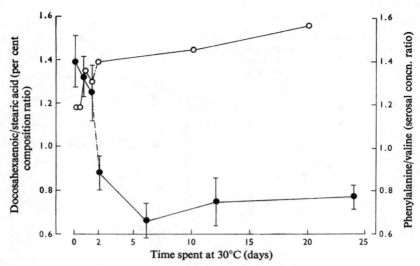

Fig. 11. *Time-course for changes in fatty acid composition of goldfish mucosal membranes and for changes in amino acid transport after a sudden increase in the environmental temperature of cold-adapted goldfish*

●, Docosahexaenoic acid/stearic acid content of intestinal membranes; ○, phenylalanine/valine serosal concentration, measured after a 2h incubation of everted intestinal sacs. Change of environmental temperature, 16°C to 30°C (Smith & Kemp, 1971).

If results found with liposomes are to apply to the biological system one might suppose that the time-course for adaptive change in membrane fatty acids would parallel that for changes in the ability of the intestine to transport amino acids at fixed concentrations. Results of experiments testing for such a correlation are shown in Fig. 11. The scatter in results has been decreased by presenting data as ratios, in one case the ratio of docosahexaenoic acid/stearic acid and in the other the ratio of phenylalanine/valine concentrations measured in the serosal fluid. Changes in the first ratio are large because the amounts of both fatty acids are changing in a reciprocal fashion. Changes in the second ratio are small because only the concentration of valine (a group 2 amino acid) is subject to regulation. What is obvious is that the changes in both ratios coincide in time. These changes, which take 48h to complete, are different in time both from the initial regulation of sodium and amino acid transport (15–20h) and from the changes in $(Na^+ + K^+)$-activated ATPase and intracellular ion concentrations (2–3 weeks). Temperature adaptation in the goldfish intestine can therefore be seen to consist of at least three different stages. These stages and their possible anatomical location within the mucosa are summarized in Fig. 12. All schemes are tentative and question marks have been included over the more debatable changes to emphasize this point.

Fig. 12(*a*) shows the mucosal cell of a 15°C-adapted fish analysed at 15°C. This cell can transport amino acids at concentrations greater than those presented to the mucosal surface and there is a small net transfer of fluid and sodium. Heating this tissue (Fig. 12*b*) causes an immediate increase in sodium influx, but the intracellular concentration falls because the rate of sodium efflux increases even more quickly. Amino acid transport also increases, the concentrations appearing

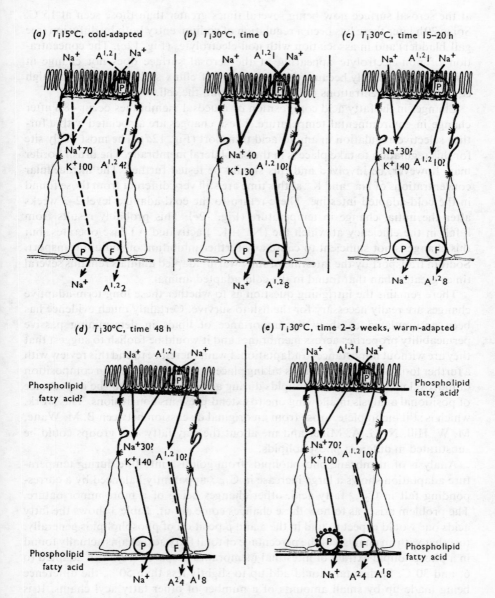

Fig. 12. *Schematic diagram of how temperature adaptation might take place in the intestinal mucosa of goldfish*

The ability to transport sodium and amino acids (A^1 for group 1 and A^2 for group 2) is indicated by the thickness of the drawn lines. The heavy horizontal lines show changes taking place in the fatty acid composition of isolated membranes. Numbers against sodium (Na^+) and potassium (K^+) give the estimated intracellular concentrations for these two ions (mM). Numbers against A^1 and A^2 give notional concentrations for typical amino acids. P and F indicate a membrane pump and a process of facilitated diffusion respectively. T_i gives the temperature at which the intestine was maintained. Question marks show where a site of adaptation is still uncertain or where the estimation of an intracellular concentration involves an element of guesswork.

at the serosal surface now being several times greater than those seen at 15°C. Some 20 h later there has been a regulation of sodium entry both by diffusion (see gall bladder) and in association with non-electrolytes (Fig. 12c). The concentration of non-electrolyte appearing at the serosal surface does not change at this time, presumably because the lower rates of efflux still allow time for high equilibrium concentrations to be built up within the cell.

Changes in the fatty acid composition of mucosal membranes occur 48 h after change in environmental temperature. These changes are associated with a further selective regulation in amino acid transport (Fig. 12d). The most likely site for this regulation to take place is in the basolateral membrane. The brush border may, however, be involved and this should be tested further. The intracellular concentrations of Na^+ and K^+ at this time are still very different from those found in the cold-adapted intestine. These return to the cold-adapted level 2–3 weeks after the initial change in temperature (Fig. 12e). This probably results from a fall in the efficiency at which the $(Na^+ + K^+)$-activated ATPase operates, but this change is not sufficient to cause any further inhibition of sodium transport. Sodium transport by the intestine of fully warm-adapted animals remains several times greater than that found in the cold-adapted animal.

There remains the intriguing question as to whether these long-term adaptive changes are really necessary for the fish to survive. Certainly much evidence has been accumulated to show the importance of lipids in determining passive permeability properties across membranes and it would be foolish to suggest that they are without importance to adaptation. I want therefore to end this review with a further look at adaptive changes taking place in the fatty acyl chain composition of mucosal membrane phospholipids during adaptation, to show the importance of positional analysis in allowing one to extend previous conclusions. This work, which is still incomplete, arose from an original discussion between B. M. Waite, M. W. Hill, N. G. A. Miller and me about the way fatty acyl groups could be substituted in different phospholipids.

Analysis of membrane phospholipids from goldfish intestine during temperature adaptation shows a large increase in $C_{18:0}$ apparently balanced by a corresponding fall in $C_{22:6}$ fatty acid, other changes being of a more minor nature. The problem arises as to how these changes come about. Table 3 shows the fatty acids one would expect to find in the α and β position of phospholipids generally, together with an analysis of the percentage of total fatty acyl chains actually found in a phospholipid extract of intestinal membranes prepared from fish adapted to 6° and 30°C. Each total should add up to slightly less than 50%, the difference being made up by small amounts of a number of other fatty acyl chains. It is obvious that the number of fatty acyl groups are too low in the α and too high in the β position of phospholipids extracted from membranes prepared from 6°C-adapted fish. This arises because the acyl groups in question ($C_{18:0}$ and $C_{22:6}$) occur on separate chains. A reciprocal change in their composition automatically vitiates any attempt to make the chains balance. Yet the accounting for total chains in the 30°C-adapted membrane phospholipids is what one might theoretically expect. The suggested answer to this paradox is that $C_{18:1}$, a fatty acid whose total percentage composition remains independent of the adaptation temperature (Table 3), switches chains, appearing in the α chain only at low adaptation

Table 3. *Fatty acid composition of membrane phospholipids showing their expected distribution between α and β chains*

Numbers give percentage of total recovered fatty acid.

Acclimatization temperature (°C) ...	α Chain			β Chain	
	6	30		6	30
$C_{16:0}$	16.5	19.1	$C_{18:1}$	12.8	11.3
$C_{18:0}$	12.1	21.2	$C_{18:2}$	6.3	11.2
			$C_{20:4}$	7.6	4.9
			$C_{20:5}$	5.5	0.7
			$C_{22:6}$	24.9	17.0
Total	28.6	40.3		57.1	45.1

Table 4. *Positional analysis of fatty acids in phosphatidylethanolamine and phosphatidylcholine isolated from intestinal membranes of warm-adapted fish (30°C)*

Numbers give the percentage of total recovered fatty acid. The recovery of $C_{18:1}$ is given for the β position, but values for the other unsaturated fatty acids in this position have been omitted for clarity.

	Phosphatidylethanolamine				Phosphatidylcholine		
	Total	α	β		Total	α	β
$C_{16:0}$	17.2	11.2		$C_{16:0}$	25.8	30.2	
$C_{18:0}$	36.9	37.5		$C_{16:2}$	2.4	2.9	
$C_{18:1}$	8.3	0	8.3	$C_{18:0}$	11.1	14.6	
				$C_{18:1}$	12.6	1.7	11.0
Total		48.7				49.4	

temperatures. To test this we have made a preliminary positional analysis of fatty acids cleaved from phosphatidylethanolamine and phosphatidylcholine using *Crotalus adamanteus* phospholipase A. The results of these experiments are shown in Table 4. The situation with phosphatidylethanolamine is clearest, with 97.4% of the α chains consisting of $C_{16:0}$ plus $C_{18:0}$. This percentage has fallen to 89.5 for phosphatidylcholine. Only 3.3% of the α chains in phosphatidylcholine and none in phosphatidylethanolamine consist of $C_{18:1}$. Virtually all the $C_{18:1}$ fatty acid is confined to the β position in the warm-adapted fish. This confirms the arrangement shown for α and β chains in Table 3 and shows why the balancing between α and β chains is satisfactory in this case.

The further prediction, which is in the process of being tested, is that $C_{18:1}$ will appear in the α chains of the cold-adapted phospholipids. If this does occur it will have further implications about membrane fluidity. Not only will the membranes lose their long-chain unsaturated fatty acyl chains on moving to a warm environment, but they will also gain $C_{18:1}$ in the β position, where it is optimally placed to carry out condensation reactions with cholesterol, causing further stabilization of the lipid bilayer.

References

Cremaschi, D., Smith, M. W. & Wooding, F. B. P. (1973) *J. Membr. Biol.* **13**, 143–164
Ellory, J. C., Nibelle, J. & Smith, M. W. (1973) *J. Physiol. (London)* **231**, 105–115
Ghosh, D., Lyman, R. L. & Tinoco, J. (1971) *Chem. Phys. Lipids* **7**, 173–184

60 M. W. SMITH

Kemp, P. & Smith, M. W. (1970) *Biochem. J.* **117**, 9–15
Klein, R. A., Moore, M. J. & Smith, M. W. (1971) *Biochim. Biophys. Acta* **233**, 420–433
Lubin, M. (1964) in *Cellular Functions of Membrane Transport* (Hoffman, J. F., ed.), pp. 193–211, Prentice Hall, Englewood Cliffs, NJ
Mepham, T. B. & Smith, M. W. (1966) *J. Physiol. (London)* **186**, 619–631
Moore, S. & Stein, W. H. (1954) *J. Biol. Chem.* **211**, 893–956
Morris, D. & Smith, M. W. (1967) *Biochem. J.* **102**, 648–653
Palay, S. L. & Karlin, L. J. (1959) *J. Biophys. Biochem. Cytol.* **5**, 363–372
Precht, H. (1958) in *Physiological Adaptation* (Prosser, C. L., ed.), pp. 50–78, American Physiological Society, Washington, DC
Schultz, S. G. & Curran, P. F. (1970) *Physiol. Rev.* **50**, 637–718
Schultz, S. G. & Zalusky, R. (1964) *J. Gen. Physiol.* **47**, 1043–1059
Smith, M. W. (1966a) *J. Physiol. (London)* **183**, 649–657
Smith, M. W. (1966b) *Experientia* **22**, 252–253
Smith, M. W. (1967) *Biochem. J.* **105**, 65–71
Smith, M. W. (1970) *Comp. Biochem. Physiol.* **35**, 387–401
Smith, M. W. & Ellory, J. C. (1971) *Comp. Biochem. Physiol. A* **39**, 209–218
Smith, M. W. & Kemp, P. (1971) *Comp. Biochem. Physiol. B* **39**, 357–365
Trier, J. S. (1968) in *Alimentary Canal* (Code, C. F. & Heidel, W., eds.), section 6, vol. 3, pp. 1125–1175, American Physiological Society, Washington, DC

Biochem. Soc. Symp. (1976) **41**, 61–109
Printed in Great Britain

Substrate Cycles in Metabolic Regulation and in Heat Generation

By ERIC A. NEWSHOLME

*Department of Biochemistry, University of Oxford, South Parks Road,
Oxford OX1 3QU, U.K.*

and

BERNARD CRABTREE

*Department of Animal Physiology and Nutrition, University of Leeds,
Vicarage Terrace, Leeds LS2 9JT, U.K.*

Synopsis

1. The presence of substrate cycles in tissues has been demonstrated by direct isotope methods in recent years. This demonstration has provided the impetus for a reappraisal of the roles of substrate cycling in metabolic regulation and in heat production. These aspects of substrate cycling are discussed in this paper. The relationship between near-equilibrium reactions and substrate cycles is emphasized, since this provides a basis for the derivation of a function describing in precise quantitative terms the factors governing the amplification provided by substrate cycles in metabolic regulation. Some examples of the roles of substrate cycles in providing sensitivity in metabolic regulation are described. The importance of substrate cycling in heat generation in the flight muscle of the bumble-bee and in brown adipose tissue is discussed in detail. 2. We point out that the two possible roles of cycling, heat production and amplification, are intimately linked so that they must be discussed together. It is proposed that variable rates of substrate cycling may be possible so that, for short periods of time, sensitivity can be maximal without excessive heat generation. Variable rates over the long term may be involved in weight control, and the control of such variability in cycling rates may be impaired in obese subjects. Finally, the possibilities that substrate cycles provide explanations for the specific dynamic action of food and for alcoholic and accidental hypothermia are raised.

A Introduction

There has been considerable increase in the knowledge of control of metabolic processes over the last 15–20 years. This advance in knowledge has stemmed from a background in basic metabolic biochemistry, which spans almost half a century and which has culminated in the production of metabolic maps that give a detailed account of the chemistry of the individual pathways that comprise metabolism. However, such maps do not provide any indication of the complex interrelated nature of metabolism. Indeed, metabolic interrelationships and metabolic control are so closely integrated that we consider that they must be considered together to provide a coherent account of regulation in metabolism.

This paper will attempt to explain how regulation in metabolism and heat generation are closely integrated.

A metabolic pathway contains reactions that are near to equilibrium and those that are far-removed from equilibrium (non-equilibrium). It is possible for a non-equilibrium reaction in the forward direction of the pathway (i.e. A → B) to be opposed by another non-equilibrium reaction which converts B into A. (Both reactions must, of course, be chemically distinct.) These two separate reactions are catalysed by enzymes E_1 and E_2 in the following hypothetical scheme

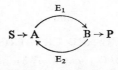

If the two enzymes E_1 and E_2 are simultaneously catalytically active, A will be converted into B and the latter reconverted back into A. This establishes a substrate cycle. (It should be noted that substrate cycles have also been termed energetically wasteful or futile cycles but, since these cycles have important metabolic functions and are not futile, we prefer the less derogatory name.)

The classical work of Schoenheimer & Rittenberg (1936) showed that even structural components (e.g. protein and carbohydrate) of an organism were not metabolically inert, but that each macromolecule must be broken down and resynthesized many times during the lifetime of the organism. The implications of this part of the work for metabolic regulation were largely ignored. Thus when two reactions such as those catalysed by E_1 and E_2 (above) occurred together in the same tissue, it was generally considered that they would never be simultaneously active, since, at best, this would lead to wastage of biological energy and at worst it could be fatal to the cell because of heat generation. Perhaps the first experimental demonstration of a metabolic substrate cycle was made by Steinberg (1963). He showed that the net rate of production of fatty acids from triglyceride in adipose tissue was often less than the rate of lipolysis as indicated by the rate of glycerol release. He concluded that some of the fatty acid produced by lipolysis was re-esterified to triglyceride: the triglyceride-fatty acid cycle was established. (This cycle is described in detail in Section D3 and Fig. 1.) However, at this time the biological significance of such cycles was not appreciated.

In 1966, Newsholme & Underwood discussed the existence of a cycle between fructose 6-phosphate and fructose diphosphate in kidney-cortex slices and proposed that the properties of the enzymes catalysing the cycle could control not only the rate, but also the direction of glucose metabolism in this tissue. The significance of substrate cycles in providing amplification in metabolic control was first proposed by Newsholme & Gevers (1967) and this concept was extended by Newsholme & Crabtree (1970), Newsholme & Start (1972), Newsholme & Crabtree (1973) and Newsholme & Start (1973, pp. 429–460).

Recently the role of substrate cycles in heat generation has attracted considerable attention. We consider that both metabolic regulation and heat generation are so intimately linked that discussion of either subject alone would be incomplete. In this paper, new information on the quantitative aspects of amplification in substrate cycles is provided, together with some details of the range of

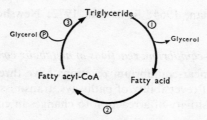

Fig. 1. *Triglyceride fatty acid cycle*

The reactions are catalysed by the following enzymes or enzyme systems: ① triglyceride lipase; ② fatty acyl-CoA synthetase; ③ enzymes of the esterification system.

metabolic processes that are regulated via such cycles. The importance of specific substrate cycles in heat generation in the bumble-bee flight muscle and the brown adipose tissue of mammals is discussed. Finally, the possibility is discussed that the concept of substrate cycles explains such physiological (and pathological) problems as the specific dynamic action of food, weight control (and obesity in the absence of this control) and alcoholic and accidental hypothermia.

Nonetheless, before such discussions can take place it is necessary to understand the terms near-equilibrium and non-equilibrium reactions in metabolic pathways and especially how near-equilibrium reactions predict some of the properties of substrate cycles.

B   Near-Equilibrium and Non-Equilibrium Reactions in Metabolic Pathways

It is known from general experience that metabolic pathways consist of near-equilibrium and non-equilibrium reactions (see Bücher & Rüssmann, 1964; Rolleston, 1972). To appreciate what is meant by these terms, it is useful to consider the rates of the forward (V_f) and reverse (V_r) processes of any reaction within a pathway. There is assumed to be a steady flux (of magnitude J) through the pathway.

$$\xrightarrow{\text{J}} A \underset{V_r}{\overset{V_f}{\rightleftharpoons}} B \longrightarrow$$

At steady state, the difference between the rates of the processes $(V_f - V_r)$ is equal to the flux, i.e. $J = V_f - V_r$. At equilibrium, $V_f = V_r$, so that $J = 0$. Consequently, in any pathway transmitting a flux, reactions cannot be at equilibrium. Nonetheless, if the ratio V_f/V_r approaches unity (<5) then the reaction is described as near-equilibrium. If this ratio is large (>5) the reaction is described as non-equilibrium (see Appendix 1).

It can be shown that the ratio V_f/V_r is quantitatively related to the displacement of the reaction from equilibrium. If K is the equilibrium constant for the reaction and Γ is the mass–action ratio (i.e. ratio of concentrations of product to concentrations of substrate at any given time), then

$$V_f/V_r = K/\Gamma \quad \text{(see Appendix 2 for derivation of this relationship)}$$

[It is important to note that one experimental method for indicating the equilibrium nature of a reaction in a tissue is the comparison of the values of K and Γ

(see Bücher & Rüssman, 1964; Rolleston, 1972; Newsholme & Start, 1973, pp. 30–32).]

1. *Significance of near-equilibrium reactions in metabolic control*

It is probable that near-equilibrium reactions have three roles in metabolic pathways: provision of reversibility of pathways, transmission of feedback control and increased sensitivity of a reaction to changes in concentrations of substrate and/or product.

(*a*) *Reversibility of metabolic pathways.* Since in near-equilibrium reactions the ratio V_f/V_r approaches unity, it is possible that, with a decrease in substrate concentration and/or an increase in that of product, the ratio can become less than unity (i.e. $V_r > V_f$). If these concentration changes are maintained (i.e. they are not merely transient), the direction of flux through the pathway has reversed. This demands that the pathway has the enzyme capability for such a reversal, which will be initiated at non-equilibrium reactions. The latter control both the rate and direction of flux through the pathway (see below). The advantage of near-equilibrium reactions is that the direction of flux can be changed merely by modulating the concentration of substrates and/or products. This represents a considerable economic advantage to the cell: non-equilibrium reactions cannot respond in such a manner and separate 'reverse-direction' reactions are required (see below). For example, in glycolysis and gluconeogenesis in liver and kidney cortex the six reactions between fructose diphosphate and phosphoenolpyruvate are near-equilibrium and serve as common reactions for both processes. However, glycolytic flux requires the specific enzymes glucokinase, phosphofructokinase and pyruvate kinase, whereas gluconeogenesis requires pyruvate carboxylase, phosphoenolpyruvate carboxykinase, fructose diphosphatase and glucose 6-phosphatase (see Fig. 2).

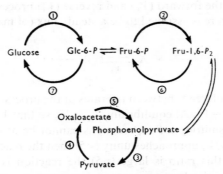

Fig. 2. *Non-equilibrium reactions of glycolysis and gluconeogenesis*

The enzymes catalysing the reactions are as follows: ① glucokinase; ② phosphofructokinase; ③ pyruvate kinase; ④ pyruvate carboxylase; ⑤ phosphoenolpyruvate carboxykinase; ⑥ fructose diphosphatase; ⑦ glucose 6-phosphatase.

(*b*) *Transmission of feedback control.* In the following hypothetical system

$$\xrightarrow{\text{E}_1} A \rightleftharpoons^{\text{E}_2} B \xrightarrow{\text{E}_3}$$

enzymes E_1, E_2 and E_3 catalyse a series of reactions in a pathway such that the product of E_1 (i.e. A) inhibits E_1. Thus any change in the rate of reaction catalysed by E_3 will be communicated to E_1 via the changes in concentrations of B and A. Since the reaction E_2 is near-equilibrium, changes in concentration of B are transmitted to A via this reaction. (It should be noted that the effect of A on E_1 is not via the reverse process of this reaction, but is an allosteric effect, i.e. A has a specific inhibitory effect on the catalytic activity of the enzyme.) An example of this control is the link that phosphoglucoisomerase provides between phospho-fructokinase and hexokinase in tissues such as brain and muscle:

Glucose $\xrightarrow{\ominus}$ glucose 6-phosphate $\rightleftarrows$ fructose 6-phosphate $\longrightarrow$ fructose diphosphate

If the conversion of glucose 6-phosphate into fructose 6-phosphate was non-equilibrium, it would be necessary for phosphoglucoisomerase to possess a specific allosteric site for inhibition by fructose 6-phosphate for changes in the activity of phosphofructokinase to be communicated to hexokinase.

(c) *Sensitivity to changes in concentrations of substrate and/or product.* We consider that the most important role of a near-equilibrium reaction in a pathway is to provide an increase in sensitivity to changes in concentrations of substrate (and/or products). Thus the reaction can accommodate large changes in flux with only small changes in concentration of these effectors. The mathematical relation-ship between the change in substrate concentration and the change in flux is derived below.

(i) Derivation of function to describe sensitivity of flux to changes in substrate concentration. A near-equilibrium reaction is considered as part of a hypo-thetical metabolic pathway (see above for reaction scheme and definition of symbols). As stated above $J = V_f - V_r$. The aim of the analysis is to find a relation-ship between the relative change in flux for a given relative change in substrate concentration (i.e. A). (It should be noted that the derivation assumes that there is no change in the concentration of B so that V_r remains constant.) Initially, the relationship between J and V_f will be derived.

It is assumed that the flux changes from J to $(J + \Delta J)$ and V_f changes to $(V_f + \Delta V_f)$. Therefore

$$J + \Delta J = (V_f + \Delta V_f) - V_r$$

Since

$$J = V_f - V_r$$

$$\Delta J = \Delta V_f$$

Therefore

$$\frac{\Delta J}{J} = \frac{\Delta V_f}{J}$$

$$= \frac{\Delta V_f}{J} \times \frac{V_f}{V_f}$$

$$= \frac{\Delta V_f}{V_f} \times \frac{V_f}{J}$$

Let the relative change in J (i.e. $\Delta J/J$) be known as $J_{rel.}$ and the relative change in V_f be known as $(V_f)_{rel.}$. Thus

$$J_{rel.} = (V_f)_{rel.} \times \frac{V_f}{J}$$

$$\frac{J_{rel.}}{(V_f)_{rel.}} = \frac{V_f}{J}$$

$$= \frac{V_f}{V_f - V_r}$$

The function V_f/J is termed the reversibility of a reaction, and is known as R [i.e. $J_{rel.}/(V_f)_{rel.} = R$]: for a completely non-equilibrium reaction the value of of R is unity (since V_r is zero, $V_f = J$), but it approaches infinity as the reaction approaches equilibrium (when $J \to 0$). The relationship between R and K/Γ is derived in Appendix 1.

If it is assumed that V_f responds linearly to a change in substrate concentration (see below for clarification) i.e. $V_f = \lambda[A]$, where λ is a constant, then the relative change in V_f equals the relative change in substrate concentration so that

$$\frac{J_{rel.}}{(V_f)_{rel.}} = \frac{J_{rel.}}{[A]_{rel.}} = R$$

Now the factor $J_{rel.}/[A]_{rel.}$ is the sensitivity of the reaction to [A], so that the sensitivity is directly proportional to R. If the reaction is non-equilibrium, so that the value of R approaches unity, there is a linear relationship between change in substrate concentration and change in flux (i.e. a 10% increase in flux will be produced by a 10% increase in substrate concentration). However, if the value of R is 10 (i.e. V_f is tenfold the flux) a 10% increase in substrate concentration will double the flux [see Newsholme & Crabtree (1973) for example]: if the value of R is 100, a 1% increase in concentration will double the flux.*

However, this effect of the reversibility R does not apply to the interaction of effectors that are not reactants, for example allosteric effectors. Such effectors alter the catalytic activity of the enzyme and therefore change both V_f and V_r such that the ratio V_f/V_r remains constant. [If $V_f = k_{+1}[A]$ and $V_r = k_{-1}[B]$, a catalytic or allosteric effector changes the values of k_{+1} and k_{-1}: since the ratio k_{+1}/k_{-1} is equal to the equilibrium constant (Appendix 2), this ratio and hence V_f/V_r cannot be changed directly by such effectors.] Applying this condition to the response of the net flux to an allosteric effector, it is found that the effect on V_r completely opposes the increased sensitivity produced by the effect on V_f: the net result is that the reversibility R has no effect on the response of J to the allosteric effector.

* It should be emphasized that in all these considerations it is assumed that the rate of the enzyme-catalysed reaction is linearly related to the change in substrate concentration (i.e. the substrate concentration is well below the K_m). The sensitivity can be improved by changing the order of reaction in relation to the substrate: for enzymes that exhibit positive co-operativity the response can be greater than linear. However, when the substrate concentration exceeds the K_m the order of reaction is less than unity so that the sensitivity decreases. This aspect of enzyme activity has been dealt with elsewhere (see Newsholme & Crabtree, 1970; Newsholme & Start, 1972, 1973, pp. 69–71; Crabtree & Newsholme, 1975).

It should be noted that the above treatment demands that concentrations of other effectors of the reaction (i.e. concentrations of product, co-substrate and co-product in the case of near-equilibrium reactions and concentrations of co-substrate and allosteric effectors in the case of non-equilibrium reactions) remain constant. For this reason the sensitivity should be termed an 'intrinsic sensitivity'.

It can be shown (by a derivation similar to that described above) that the intrinsic sensitivity to the product of the reaction (i.e. $J_{rel.}/[B]_{rel.}$) is given by $-(R-1)$, i.e. $J_{rel.}/[B]_{rel.} = -(R-1)$. The negative sign indicates that increased product concentration decreases the net flux J, although the absolute value of the sensitivity still increases with the value of R.

The effort taken to explain quantitatively the relationship between reversibility of a reaction and sensitivity has been undertaken because it provides the basis for considering the sensitivity provided by substrate cycles. (This is discussed below.) If a non-equilibrium reaction in a pathway is 'reversed' by an opposing non-equilibrium reaction, this provides a degree of reversibility of the overall step, so that consideration of the amplification provided by near-equilibrium reactions provides a logical antecedent to a discussion of substrate cycles and amplification.

2. *Significance of non-equilibrium reactions in metabolic control*

One important difference between a near-equilibrium and a non-equilibrium reaction is that in the latter the rate of the reverse reaction is insignificant in relation to the forward reaction, i.e. $(V_f \gg V_r)$, so that the product concentration has no significant effect on the flux through the reaction (at least, via mass–action effects). The significance of the non-equilibrium reactions in metabolic control has been discussed elsewhere (Newsholme & Start, 1973, pp. 15–27; Crabtree & Newsholme, 1975), so that it will be summarized only briefly here. Non-equilibrium reactions provide directionality to a pathway. Any increase in catalytic ability of the enzyme will increase the rate of the reaction in the direction of product formation. This increase in rate will be caused either by an increase in concentration of substrate (or co-substrate) or that of an allosteric regulator. If under some physiological conditions the direction of flux has to be reversed, this can only be done by the presence of an enzyme that catalyses a reverse reaction to that of the forward reaction, although to avoid violation of the laws of thermo-dynamics the chemical nature of the reactions must be different. The presence of both enzymes provides the potential for a substrate cycle. One disadvantage of a non-equilibrium reaction is that the sensitivity is small (see above). The sensitivity can be increased either by changes in the response of the enzyme to changes in substrate or regulator concentrations (i.e. positive co-operativity; Newsholme & Crabtree, 1970; Newsholme & Start, 1972, 1973, pp. 69–71; Crabtree & Newsholme, 1975) or by development of a substrate cycle. The latter is discussed in detail below.

C Substrate Cycles in Metabolic Regulation

A metabolic pathway is a series of reactions through which the substrate of the pathway is converted into the product. Some of the reactions of the pathway are near-equilibrium whereas others are non-equilibrium (see Section B). The significance of the non-equilibrium reactions is to provide directionality in the pathway. However, if it is necessary to reverse the direction of flux in this pathway (i.e. convert product into substrate), the non-equilibrium reactions provide kinetic and thermodynamic barriers. These problems are overcome by the presence of enzymes that catalyse separate 'reverse' reactions which 'oppose' the forward direction of the original pathway. (Excellent examples of this situation are provided by reference to the non-equilibrium reactions of glycolysis and gluconeogenesis in liver and kidney cortex.) If the two sets of 'opposing' reactions are found in the same compartment of the cell, the problem arises as to how the enzymes are controlled so that when the flux is in one direction, the 'reverse' enzyme is totally inhibited and *vice versa*. However, we consider that such a situation is very unlikely, if not impossible; the efficacy of metabolic control cannot be such as to inhibit totally the activity of an enzyme (unless the enzyme protein is destroyed and its synthesis prevented). Consequently, when two such 'opposing' enzymes exist in the same compartment they will both be simultaneously active and catalyse a substrate cycle. (The activities of both enzymes can be changed by metabolic regulators and the enzymes may be so regulated that the changes occur in opposite directions.) The historical development of the concept of substrate cycles in metabolic regulation has been described in Section A.

1. *Definition of a substrate cycle*

A substrate cycle is produced by two non-equivalent fluxes: one flux represents the flow of metabolites through the cycle, the other represents the flow of metabolites around the cycle. In the following cycle

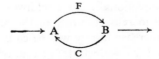

the magnitude of the former flux equals $(F-C)$ whereas that of the latter flux equals C. Thus the substrate cycle consists of a cyclic flux which must be coupled to (i.e. driven by) another flux.* The cycle need not be restricted to a set of simple opposing reactions as above, but may involve several reactions and it may be more complex so that it extends over two or more tissues (see below). Examples of substrate cycles in metabolism will be described in Section D, in which the significance of the cycles in metabolic control will be discussed in detail. The role of substrate cycles in heat generation will be described in Section E.

In the rest of this section the analogies between substrate cycles and near-equilibrium reactions will be indicated and this will lead to the derivation of functions defining quantitatively the increased sensitivity provided by cycles in metabolic control.

---

* In the substrate cycle between fructose 6-phosphate and fructose diphosphate (Section D1), the cyclic flux is driven by the hydrolysis of ATP.

2. *Substrate cycles and near-equilibrium reactions*

If two metabolites are linked via a substrate cycle, the system has similarities to a near-equilibrium reaction. In both cases there is a reverse reaction, but in the substrate cycle this is a different reaction from the forward one. Furthermore, since the substrate cycle comprises, at least, two different non-equilibrium reactions it must be provided with a source of free energy. In contrast, a near-equilibrium reaction does not require an input of energy: the reverse process is the exact thermodynamic reversal of the forward process

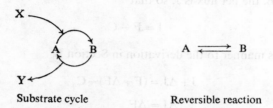

Substrate cycle   Reversible reaction

(In the substrate-cycle example, the presence of X and Y in the reactions indicates the non-equilibrium nature of each reaction.) These differences between cycles and near-equilibrium reactions are reflected in their regulatory properties (see below).

(*a*) *Reversal of flux.* As already indicated, the presence of a substrate cycle provides the potential for reversing the direction of flux in the pathway. Thus the property of the cycle is somewhat similar to that of a near-equilibrium reaction. However, there is an important difference. Since the regulator that causes the reversal in flux is modifying a non-equilibrium reaction in the cycle, the regulator can be either a mass–action effector (i.e. substrate or co-substrate) or an allosteric effector. In a near-equilibrium reaction only changes in concentration of substrates or products (i.e. mass–action effectors) can produce reversal of flux.

(*b*) *Provision of a feedback link.* In a similar manner to the near-equilibrium reaction, a substrate cycle can provide a link between two non-equilibrium reactions that are separated within a metabolic pathway by the cycle. In the following hypothetical system

$$E_1 \xrightarrow{\quad} A \underset{E_R}{\overset{E_F}{\rightleftharpoons}} B \xrightarrow{\quad E_2 \quad}$$

a change in the activity of E_2 can be communicated to E_1 via the change in the concentration of A which is influenced by the mass–action effect of B on the reaction catalysed by enzyme E_R. Another, perhaps more subtle, manner in which a substrate cycle plays a role in feedback regulation of fatty acid mobilization in adipose tissue will be discussed in Section D.

(*c*) *Substrate cycles and sensitivity in metabolic control.* The derivation of a function to describe the sensitivity provided by a substrate cycle is somewhat analogous to that for the near-equilibrium reaction described in Section B.

The substrate cycle is described diagrammatically as follows

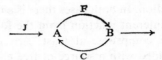

where F and C describe the rates of the non-equilibrium reactions that inter-convert A and B, the net flux is J, so that

$$J = F - C$$

In an analogous manner to the derivation in Section B,

$$J + \Delta J = (F + \Delta F) - C$$

$$\therefore \Delta J = \Delta F$$

$$\frac{\Delta J}{J} = \frac{\Delta F}{J} = \frac{\Delta F}{F} \times \frac{F}{J}$$

$\Delta J/J$ is the relative increase in flux and it is known as $J_{rel.}$, $\Delta F/F$ is known as $F_{rel.}$. Therefore (as in Section B)

$$J_{rel.} = F_{rel.} \times \frac{F}{J}$$

so that

$$\frac{J_{rel.}}{F_{rel.}} = \frac{F}{J}$$

Since $J = F - C,\ F = J + C$ so that

$$\frac{J_{rel.}}{F_{rel.}} = \frac{J + C}{J} = 1 + \frac{C}{J}$$

If it is assumed that F is linearly related to the concentration of substrate A (i.e. $F = \lambda[A]$, where λ is a constant) so that $F_{rel.} = [A]_{rel.}$ (in other words the concentration of substrate is well below the K_m of the enzyme for that substrate), it follows that

$$\frac{J_{r.el}}{[A]_{rel.}} = 1 + \frac{C}{J}$$

Thus the sensitivity of the reaction to [A], $J_{rel.}/[A]_{rel.}$ is equal to $1 + C/J$. Since C is equivalent to the rate of cycling and J is the flux through the reaction, the sensitivity of the reaction to changes in substrate concentration is dependent on the ratio, cycling rate/flux. This was first pointed out by Newsholme & Start (1972) and has now been put on a precise quantitative basis.

It should be stressed that this sensitivity function refers not only to a substrate, but also to allosteric regulators of the reaction F. (This is an important difference from near-equilibrium reactions and results from the thermodynamic independence of the two opposing reactions.) In this case, the function will be

$$\frac{J_{rel.}}{[X]_{rel.}} = 1 + \frac{C}{J}$$

where X is the allosteric regulator of the forward reaction.

Similar functions can be derived for the effects of substrate or allosteric regulators on the reverse reaction (i.e. B → A) in the cycle. It can be shown (see Appendix 3) that

$$\frac{J_{rel.}}{[B]_{rel.}} = -\frac{C}{J}$$

(The negative sign indicates that an increase in the concentration of B decreases the flux.)

The sensitivity of a non-equilibrium reaction (i.e. R = 1), in the absence of any cycling, to change in concentration of substrate or allosteric effector that interacts linearly is given by the equation (see Section B1c):

$$\frac{J_{rel.}}{[X]_{rel.}} = R$$

where R approaches unity. Thus the additional sensitivity provided by the substrate cycle increases with the factor cycling rate/fluux. It should be noted that since this ratio is added to unity (i.e. sensitivity = $1+C/J$), the ratio of cycling to flux must be large to provide a marked increase in sensitivity to effectors of the forward reaction. This has been discussed by Newsholme & Start (1972) and by Newsholme & Crabtree (1973) and its implications are discussed in some detail in Section F.

The work of D. G. Clark et al. (1973) on the changes in isotopic content of [^{14}C, ^{3}H]glucose in rat hepatocytes incubated in vitro has provided direct evidence for the existence of a substrate cycle between glucose and glucose 6-phosphate catalysed by glucokinase and glucose 6-phosphatase. Furthermore, from the data provided by these workers it is possible to calculate the rates of the forward and reverse reactions in the substrate cycle. This has been done and the results are reported in Table 1.

From these results it can be seen that the rate of substrate cycling in comparison with the flux is not large and the values of the function $1+C/J$ are 3.3 and 1.75 for

Table 1. *The sensitivity provided by the substrate cycle between glucose and glucose 6-phosphate in intact liver cells*

Data in the Table are compiled from those of D. G. Clark et al. (1973). J, C and F are defined in the text.

Condition of rat	Presence of lactate	Glucose formation (μmol/h per 100 mg of cells) (J)	Glucokinase activity (C)	Glucose 6-phosphatase activity (F)	1+C/J (in absence of lactate)
Starved	−	2.3	5.2	7.5	3.3
	+	27.3	14.0	41.3	
Fed	−	12	9.0	21.0	1.75
	+	32	10.0	42.0	

the 'starved' and 'fed' hepatocytes respectively. Thus the cycle has increased the sensitivity (i.e. $J_{rel}/[Glc-6-P]_{rel}$) by a factor of 3.3- and 1.75-fold respectively. It should be noted that in all situations reported there is a significant rate of glucose formation (i.e. flux is large). If the flux rate was small (e.g. 0.1 μmol/h), the sensitivity would be very large (up to 141).

D   Role of Some Substrate Cycles in Metabolic Regulation

The potential number of substrate cycles is fairly large, so that it is not possible to discuss the possible roles of all cycles in metabolic regulation in this paper. The cycles that we have selected for discussion represent those for which there is some direct evidence or which provide examples of important principles of the role of cycles in metabolic regulation.

1.   Substrate cycle between fructose 6-phosphate and fructose diphosphate

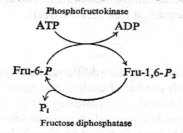

This cycle is catalysed by the activities of phosphofructokinase and fructose diphosphatase. There is evidence that it is present in liver, kidney cortex and some muscles [see Katz & Rognstad (1975) for review]. In the two former tissues, fructose diphosphatase is present to provide one of the non-equilibrium reactions of the gluconeogenic pathway, which occurs when the glycolytic flux is reversed.

The cycle has historic importance in kidney, since results on the control of the rates of glycolysis and gluconeogenesis with kidney-cortex slices were first explained on the basis of a substrate cycle at this level (Newsholme & Underwood, 1966; Underwood & Newsholme, 1967). This led to the concept of amplification discussed above (Section C). In recent years, direct isotopic evidence for the operation of this cycle in liver has been obtained (D. G. Clark et al., 1973) and it has been suggested that the hormone glucagon modulates the rate of glucose formation, at least in part, through effects on this cycle (Clark et al., 1974).

In muscle, the distribution and properties of the enzyme fructose diphosphatase led to the idea of this substrate cycle playing a role in the amplification in the control of glycolysis. This concept has been discussed in detail elsewhere and it will not be elaborated further (Newsholme & Crabtree, 1970; Newsholme, 1972; Newsholme & Crabtree, 1973; Newsholme & Start, 1973, 121–124). However, we would like to emphasize that the increased sensitivity of glycolytic flux to changes in the concentration of AMP, which is conferred by the cycle, provides a direct regulatory link between the ATP-utilizing process (i.e. contraction) and a main ATP-producing system (i.e. glycolysis). We consider that this ensures that the glycolytic flux provides precisely the rate of ATP formation required by the muscles. Thus an increased rate of ATP utilization decreases the concentration of ATP and increases that of AMP until the glycolytic flux provides ATP at a rate equal to that of ATP utilization, at which point the concentrations of ATP and AMP become steady. Such a degree of precision could not be obtained solely by regulation through regulatory factors external to the rate of ATP utilization by the muscle (e.g. changes in Ca^{2+} or cyclic AMP concentrations).

At this stage, it should be pointed out that it is necessary to modify the substrate-cycling concept at this level in muscle in order to take into account the amounts of heat generated by the cycle (see Section F).

2. Substrate cycle between glucose and glucose 6-phosphate in liver

The cycle is catalysed by glucokinase and glucose 6-phosphatase,

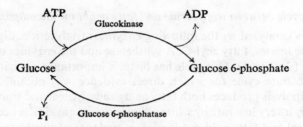

The role of the cycle in the control of the blood glucose concentration was first discussed by Newsholme & Gevers (1967) and the concept was extended further by Newsholme & Start (1973, pp. 267–269). Direct evidence has been obtained for the operation of this cycle in liver (D. G. Clark *et al.*, 1973). The role proposed for the cycle is that changes in the blood glucose concentration in the hepatic portal vein (and thus in the concentration of intracellular glucose) control the rates of both glucose uptake and glucose release. Thus fairly small changes in blood glucose concentration can have large effects on the rate and direction of glucose metabolism in the liver. This sensitivity is provided by the existence of the substrate cycle [see Newsholme & Start (1973, pp. 267–269) for details]. Further, it has been suggested that impaired glucose tolerance in patients suffering from liver disease (e.g. hepatitis, cirrhosis) may be due to a low rate of this substrate cycle. Consequently, the sensitivity of the glucose uptake and release processes to changes in the concentrations of hepatic portal blood glucose may be markedly decreased in these patients (see Newsholme, 1976).

It is also possible that the rate and direction of glucose metabolism through the cycle are modified by changes in the concentration of glucose 6-phosphate, since the hepatic concentration of the latter is similar to the K_m of glucose 6-phosphatase (see Hue & Hers, 1974). Thus an increase in the intracellular level of glucose 6-phosphate could increase the rate of glucose production or decrease the rate of glucose uptake by the liver. Since glucose stimulates allosterically the activity of hepatic UDP-glucose glucosyltransferase, Hue & Hers (1974) have suggested that an increase in portal blood glucose will lead to a stimulation of glycogen synthesis, which will lower the intracellular concentration of glucose 6-phosphate and decrease the activity of glucose 6-phosphatase. This will increase the net rate of glucose phosphorylation, so that the uptake of glucose by the liver is increased. The overall effect is to assist the conversion of blood glucose into liver glycogen.

It should be pointed out that if both enzymes of the cycle are sensitive to their respective substrates, the effect of increasing the glucose concentration (and glucokinase activity) could be negated by the increase in the concentration of glucose 6-phosphate (which would increase the activity of the phosphatase). In the direction of glucose utilization, this can be minimized by the independent increase in activity of either UDP-glucose glucosyltransferase or phosphofructo-kinase (the transferase activity is stimulated by glucose and glucose 6-phosphate and phosphofructokinase is stimulated by fructose 6-phosphate). These effects would minimize changes in the concentration of glucose 6-phosphate. In the direction of glucose production, opposing effects due to the increase in glucose concentration may be minimized by the removal of glucose via the bloodstream.

3. *Substrate cycle between triglyceride and fatty acid in white adipose tissue*

This cycle is catalysed by the following enzymes: triglyceride, diglyceride and monoglyceride lipases, fatty acyl-CoA synthetase and the enzymes of the esterification system (see Fig. 1). This cycle has historic importance since it was the first metabolic substrate cycle for which direct evidence was obtained (Steinberg, 1963). Since lipolysis produces both fatty acids and glycerol and since the latter is utilized at only a very low rate by white adipose tissue, any difference between the rates of glycerol and fatty acid production is evidence of substrate cycling. This was first demonstrated by Steinberg (1963).

Two roles have been suggested for this cycle. First it links the changes in plasma glucose concentration with rates of fatty acid mobilization from adipose tissue (see Newsholme & Start, 1973, pp. 217–218). Second, as with other cycles, it provides a more sensitive system for the regulation of the rate of fatty acid mobilization from adipose tissue. Although this regulation is considered to be effected largely through the effects of various hodmones [see Steinberg (1963); for review, see Newsholme & Start (1973), pp. 221–224], the rate of mobilization of fatty acids from adipose tissue must be precisely controlled in relation to the rate of utilization of fatty acid for energy production by the tissues, especially muscle tissue. If the rate of mobilization is too low, the muscle will not receive sufficient fatty acid for energy production so that glucose will be used. Since glucose stores in the body are very limited, this will soon result in exhaustion. If the rate of fatty acid release is too high, the plasma fatty acid concentration will increase to values that could be toxic. Consequently, a third role for the triglyceride–fatty acid cycle has been suggested: the feedback regulation of fatty acid mobilization from adipose tissue.

(*a*) *Feedback regulation of mobilization of fatty acids from adipose tissue.* The significance of fatty acid mobilization from adipose tissue is to supply a fuel for oxidation in other tissues (especially muscle). This has been realized since the classic work of Gordon & Cherkes (1956) and Havel & Fredrickson (1956). However, two aspects of the mechanism by which fatty acid mobilization is regulated have been neglected. First, adipose tissue is dispersed throughout the animal (it is not present as a discrete tissue such as liver), so that a precise control of the rates of overall mobilization of fatty acids might be physiologically difficult.

Second, the rate of fatty acid mobilization must be exactly sufficient to meet the energy needs of the animal. These needs can be very variable.

The triglyceride–fatty acid cycle in adipose tissue provides a subtle mechanism for feedback regulation of the rate of fatty acid mobilization, so that a fairly constant concentration of fatty acid is maintained in the blood despite fluctuations in the rate of utilization. If the fatty acid concentration in the blood decreases, owing to increased peripheral utilization (i.e. increased muscle activity), this will lower the concentration of fatty acids in the adipose-tissue cell and decrease the rate of esterification (since the esterification pathway is not saturated with fatty acyl-CoA). Consequently, a greater proportion of the fatty acids produced by lipolysis will enter the blood from the adipose tissue and thus satisfy the increased peripheral demand. Similarly, if the rate of mobilization is too high, the blood fatty acid concentration will increase and stimulate the rate of esterification. Consequently, fatty acid mobilization from adipose tissue is decreased (see Fig. 3). Thus this cycle in adipose tissue performs a similar role to the fructose 6-phosphate–fructose diphosphate cycle in muscle: the cycles provide intracellular regulation of high precision between supply and demand of energy or fuel in response to less precise external regulation (e.g. hormones, nervous stimuli).

[It should be noted that at least one other feedback control exists in adipose tissue to adjust the rate of fatty acid mobilization to the energy needs of the animal: this is the antilipolytic action of ketone bodies that may be either direct or indirect; see Newsholme & Start (1973), pp. 310–312 and Newsholme (1976) for reviews.]

(b) *Effects of adrenaline on the cycle.* It is well established that adrenaline increases the plasma fatty acid concentration due to increased lipolysis in adipose tissue (see Scow & Chernick, 1970). However, it is also reported that adrenaline increases the rate of both lipolysis and esterification so that the rate of cycling is increased (Steinberg, 1963; Denton *et al.*, 1966). It is suggested that this increased rate of cycling increases (at least transiently) the ratio cycling rate/flux. We suggest that this increase in substrate-cycling rate increases the sensitivity of the process of fatty acid mobilization in anticipation of a stress situation. Thus a subsequent relatively small decrease in the blood fatty acid concentration due to increased peripheral utilization (i.e. due to a large increase in contractile activity, the 'flight or fight' activity) will decrease the rate of esterification and cause a marked increase in the rate of mobilization of fatty acids. This is an example of the role of adrena-

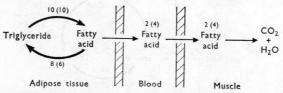

Fig. 3. *Feedback control of fatty acid mobilization from adipose tissue via the triglyceride–fatty acid cycle*

The numbers represent the hypothetical rates of the various processes. The numbers in parentheses indicate the changes in rates produced by an increased rate of fatty acid utilization by muscle.

line in stimulating the rate of substrate cycles to increase their sensitivity for short periods of time. This is discussed in detail in Section F.

4. *Substrate cycle between triglyceride and fatty acid in muscle*

This is the same cycle as occurs in adipose tissue, but in muscle, any net production of fatty acids by the cycle is not released from the tissue but is oxidized to CO_2 and water. Thus in some muscles it has been shown that fatty acids derived from endogenous triglyceride can be oxidized to provide energy for contraction (see Denton & Randle, 1967). However, it seems likely that, in most situations, sufficient quantities of fatty acid from adipose tissue are provided via the bloodstream for the requirements of the muscle (Carlson, 1969) so that endogenous triglyceride is not broken down (Nasoro *et al.*, 1966). Furthermore, in many muscles the activities of triglyceride lipase are much lower than those of the fatty acid oxidation enzymes (see Crabtree & Newsholme, 1972, 1975) so that the major role of the triglyceride–fatty acid cycle may not be the provision of fatty acids for oxidation. We suggest that the role of the cycle in muscle is to buffer the intracellular concentration of fatty acids (see Fig. 4). An increase in the intracellular concentration of fatty acids increases the rate of oxidation in muscle, but the rate of oxidation must be strictly dependent on the rate of ATP utilization by the muscle. Thus if the blood fatty acid concentration is elevated and some muscles are not mechanically active, the intracellular concentration in these muscle fibres could increase to very high values, since the fatty acids would not be oxidized. In some conditions, they might increase to values that cause intracellular damage. However, there is a metabolic fate for fatty acids other than oxidation: this is esterification. An increase in the rate of conversion of the fatty acids into triglyceride could prevent excessive accumulation of fatty acids. The cycle would provide sensitivity to small changes in the intracellular concentration of fatty acids, as described above (Section C), so that small increases in their intracellular concentration would stimulate esterification and hence their removal.

This hypothesis provides an explanation for the increased content of triglyceride in muscle in alloxan-diabetic animals (see Randle *et al.*, 1966). The very high concentrations of fatty acids in the blood in this condition may increase the

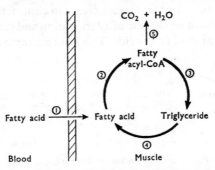

Fig. 4. *Triglyceride–fatty acid cycle in muscle*

The reactions are as follows: ① transport of fatty acid across the cell membrane; ② fatty acyl-CoA synthetase; ③ esterification reactions; ④ lipolysis; ⑤ oxidation.

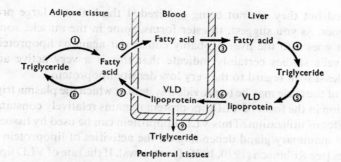

Fig. 5. *Triglyceride–fatty acid cycle between adipose tissue and liver*

The reactions are as follows: ① lipolysis; ② transport of fatty acid out of the adipose tissue; ③ fatty acid uptake by liver; ④ esterification; ⑤ formation of very-low-density lipoprotein from triglyceride; ⑥ transport of VLD lipoprotein into blood; ⑦ hydrolysis of the triglyceride in VLD lipoprotein to fatty acids which are taken up by adipose tissue: ⑧ esterification; ⑨ utilization of the triglyceride in VLD lipoprotein by peripheral tissues.

intracellular concentration and increase the rate of the esterification pathway, despite the fact that some of these muscles will be oxidizing fatty acids.

5. Substrate cycle between triglyceride and fatty acid between adipose tissue and liver

A high proportion of the fatty acids released by the adipose tissue can be taken up by the liver and esterified to form triglyceride (Robinson, 1970). These triglycerides are released by the liver into the bloodstream in the form of very-low-density lipoproteins (VLD lipoproteins). They are subsequently removed as fatty acids by the adipose tissue, where they are re-esterified. Thus a fatty acid triglyceride cycle exists between these two tissues (see Fig. 5). For convenience, the cycle can be truncated and considered as a cycle between triglyceride in the adipose tissue and triglyceride in the blood (see Fig. 6). This cycle was described by Steinberg (1963) as follows. 'When we stimulate with catecholamine and get a 20% increase in oxygen consumption, the increase in the turnover of the fatty acid is 3 or 4 times as great. If we calculate how much oxygen would be needed to oxidize this extra fatty acid that is being turned over, it is far in excess of the extra oxygen the patient actually uses. In other words, a great deal of this increased turnover is fuss and feathers and getting nowhere. The fatty acids are

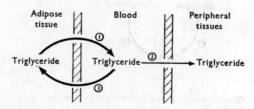

Fig. 6. *Truncated version of the intertissue triglyceride fatty acid cycle*

The reactions comprise those in Fig. 5: ① is the release of fatty acids from adipose tissue and the formation of triglyceride for the blood; ② utilization of triglyceride by peripheral tissues; ③ uptake of triglyceride fatty acids and their esterification by adipose tissue.

78 E. A. NEWSHOLME AND B. CRABTREE

being moved but they are not being oxidized. I think that a large proportion probably goes, as you suggest, to ester forms, some in the muscle, some in the liver. What goes into the liver probably comes out again as lipoprotein triglyceride. Havel's studies certainly indicate that this is a very active and rapid system, at least with regard to the very low density lipoprotein.'

The role of the cycle may be to provide a system in which the plasma triglyceride concentration in the form of VLD lipoprotein remains relatively constant despite different rates of utilization. Thus VLD lipoprotein can be used by tissues such as muscle and mammary gland depending on the activities of lipoprotein lipase in these tissues [see Robinson (1970, 1976) for reviews]. If the rate of VLD lipoprotein utilization by muscle increases, the resultant fall in its blood concentration could decrease the uptake by the adipose tissue and thus increase the net flux of VLD lipoprotein production. Consequently, more triglyceride is being 'directed' towards the tissue that is utilizing triglyceride at a greater rate.

The role of this cycle is not unlike the previous examples in that it helps to maintain the concentration of a fuel despite large variations in its rate of utilization.

6. Substrate cycle between hexose monophosphates and glycogen

The enzymes involved in this cycle are phosphorylase, UDP-glucose glucosyltransferase and pyrophosphorylase. Since glucose 1-phosphate is in equilibrium (or near-equilibrium) with glucose 6-phosphate and fructose 6-phosphate, the cycle involves hexose phosphates rather than only glucose 1-phosphate (see Fig. 7). The cycle occurs in both liver and muscle and it is useful to discuss it in these two tissues separately.

(a) The cycle in liver. The role of the cycle in liver may be to stabilize the concentration of hexose monophosphates despite large changes in flux and/or direction of glucose metabolism. In this tissue, the glucosyltransferase activity is increased both by glucose and glucose 6-phosphate (Hue & Hers, 1974; Steiner & King, 1964). The cycle provides a more sensitive system to changes in the concentration of these regulators (see Section D2).

An increase in the portal blood glucose concentration leads to an increase in the hepatic glucose concentration and this results in a stimulation of glycogen synthesis, whereas the concomitant changes in glucose 6-phosphate concentration

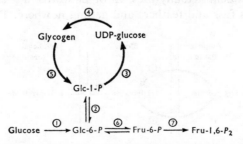

Fig. 7. Substrate cycle between hexose monophosphates and glycogen

The reactions are as follows: ① hexokinase or glucokinase; ② phosphoglucomutase; ③ UDP-glucose pyrophosphorylase; ④ UDP-glucose glucosyltransferase; ⑤ phosphorylase; ⑥ phosphoglucoisomerase; ⑦ phosphofructokinase.

are small (see Newsholme & Start, 1972). Conversely, a lowering of blood glucose causes a breakdown of liver glycogen although the glucose 6-phosphate concentration is maintained. It is well established that the breakdown of liver glycogen maintains the blood glucose concentration remarkably constant during the diurnal starvation. Start & Newsholme (1968) demonstrated that the hepatic hexose monophosphate concentrations remain constant during the breakdown of glycogen; when almost all the glycogen has been degraded (i.e. after about 24h of starvation), the concentration of hexose monophosphates falls dramatically. Thus it is suggested that the cycle between glycogen and hexose phosphates provides a system in which small changes in liver glucose and glucose 6-phosphate concentrations modulate the activity of glucosyltransferase, so that the rate of glycogen breakdown is adjusted precisely to the demand for glucose by the whole animal during the early stages of starvation (and particularly during the diurnal starvation). In the fed animal, the same properties of the enzyme also permit precise control of net glycogen synthesis during carbohydrate absorption from the gut. It is suggested that the properties of one enzyme (i.e. UDP-glucose glucosyltransferase) provide control of the flux in both directions through the operation of the cycle.

It has also been pointed out that maintenance of the hexose monophosphate concentrations in the liver during glycogen breakdown will minimize the rate of hexose monophosphate formation from fructose diphosphate, so that the rate of gluconeogenesis will be restrained. (This occurs because the normal concentration of fructose 6-phosphate in the liver maintains the activity of phosphofructokinase and thus the rate of substrate cycling between fructose 6-phosphate and fructose diphosphate.) Once the glycogen stores are depleted, the concentration of the hexose monophosphates is decreased and this lowers the activity of phosphofructokinase. An increase in the net rate of formation of fructose 6-phosphate from fructose diphosphate occurs because of the substrate cycle (see Start & Newsholme, 1970). This hypothesis may explain part of the biochemical mechanism by which gluconeogenesis is 'switched on' when the glycogen stores in the liver are depleted.

(b) *The cycle in muscle.* It is possible that a similar cycle exists between glycogen and glucose 1-phosphate in resting muscle. However, we consider that in this tissue changes in the activity of phosphorylase via changes in the concentrations of its allosteric regulators (e.g. AMP, ATP, glucose 6-phosphate, glucose 1-phosphate) may be more important in controlling the rate of cycling than changes in the activity of glucosyltransferase. Large changes in the activity of phosphorylase are produced via the interconversion cycle catalysed by phosphorylase *b* kinase and phosphorylase *a* phosphatase, which is probably the most important control mechanism of the phosphorylase enzyme during changes in mechanical activity in the muscle. If the major physiological control of phosphorylase activity is the conversion of phosphorylase *b* into phosphorylase *a*, the question arises as to the significance of the substrate cycle between glycogen and glucose 1-phosphate. Its role may be to decrease the net flux from glycogen to glucose 1-phosphate so that it is almost zero under resting conditions. Since resting muscle uses very little ATP in comparison with the ATP consumption when it is mechanically active, it is important that glycogen is not slowly degraded during the periods of rest, which in some periods may be prolonged. However, it is also possible that the rate of sub-strate cycling between glycogen and glucose 1-phosphate in resting muscle may be stimulated in stress situations by the effect of adrenaline. The latter increases the activity of phosphorylase *b* kinase via effects of 3':5'-cyclic AMP on the protein kinase. Phosphorylase *b* kinase increases the rate of conversion of less active phosphorylase *b* into more active phosphorylase *a* [for reviews see Fischer *et al.* (1970), Newsholme & Start (1973, pp. 158–185)].

Adrenaline is released in stress situations before muscular activity ('fight or flight'). Thus, although phosphorylase *b* kinase may be activated and an increased proportion of phosphorylase *a* produced, glycogen breakdown to lactate will be restricted owing to lack of demand for ATP, since the muscle is at rest. It is suggested that adrenaline may increase transiently the rate of two cycles: the interconversion cycle between phosphorylase *a* and phosphorylase *b* and the sub-strate cycle between glycogen and hexose monophosphates. Thus one further example of adrenaline increasing (transiently) the rate of substrate cycling (and interconversion cycling) is provided. This effect of adrenaline increases the ratio cycling rate/flux, and thus increases the sensitivity of the flux to other metabolic controls. Since the interconversion cycle is regulated by changes in [Ca^{2+}], it is suggested that the effect of adrenaline is to sensitize the phosphorylase inter-conversion cycle to changes in this ion. Therefore a large increase in the rate of formation of phosphorylase *a* will occur on initiation of contraction. Similarly, an increase in the rate of substrate cycling between glycogen and hexose mono-phosphate will increase the sensitivity of glycogen degradation to the allosteric regulators of phosphorylase *b* (see above) and the *b* into *a* conversion.

7. *Substrate cycle between nicotinamide nucleotides in mitochondria*

The transfer of reducing equivalents between mitochondrial NAD^+ and $NADP^+$ is catalysed by two distinct enzymes or enzyme systems. One trans-hydrogenase, referred to as energy-linked, is coupled to the operation of the respiratory chain; the other, referred to as non-energy-linked, is independent of the respiratory chain. For example, glutamate dehydrogenase catalyses a near-equilibrium reaction in the mitochondria but reacts with both NAD^+ and $NADP^+$. Consequently, this acts as a non-energy-linked transhydrogenase (see Krebs & Veech, 1970). The combination of two such systems provides a substrate cycle as follows:

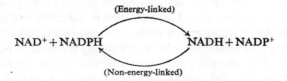

$$\text{NAD}^+ + \text{NADPH} \quad \overset{\text{(Energy-linked)}}{\underset{\text{(Non-energy-linked)}}{\rightleftarrows}} \quad \text{NADH} + \text{NADP}^+$$

Rydström *et al.* (1970) have obtained evidence for a cycling of reducing equivalents between the nicotinamide nucleotides as indicated above. The role of this cycle may be to provide a feedback mechanism by which, for example, an increase in the rate of oxidation of NADH via the electron-transport chain increases the net flux of reducing equivalents from NADPH to NAD^+. In this case the $NADPH/NADP^+$ system in the mitochondria could act as a buffer system to maintain the constancy of the $NADH/NAD^+$ system, despite variations in the rate of oxidation of NADH. To this extent, the system may be similar to the role of creatine phosphate in buffering the ATP concentration in the sarcoplasm of muscle. The presence of the cycle may also increase the sensitivity of the net flux to changes in concentration of NAD^+ and NADH. It is interesting that the presence of this cycle overcomes the problems of the lack of specificity of glutamate dehydrogenase and it overcomes the criticism that this enzyme cannot catalyse a near-equilibrium reaction because of the loss of energy when it is considered in relation to energy-linked transhydrogenation. The substrate cycle loses energy to provide sensitive regulation.

E Substrate Cycles and Heat Generation

The importance of substrate cycles in metabolic regulation has been emphasized in Section D. The production of heat is an unavoidable consequence of substrate cycling, so that the two products of these cycles are amplification and heat generation. The problem of the amount of heat generated by substrate cycles in metabolic control will be considered in Section F. However, it is possible to consider that certain substrate cycles have been developed primarily for the production of heat. There is now considerable evidence that specific heat-production mechanisms occur in brown adipose tissue and in the flight muscle of the *Bombus* species of bumble-bee (see Smith & Horwitz, 1969; Horwitz & Smith, 1972; Heinrich, 1972a,b; Newsholme et al., 1972). The involvement of substrate cycles in the generation of heat in these tissues will be considered below. The following discussion will also contain some comment on the amount of heat generated by substrate cycling. Details of how the amount of heat production can be calculated will be deferred until Section F.

1. Production of heat by the fructose 6-phosphate–fructose diphosphate substrate cycle in flight muscle of the bumble-bee

In the flight muscle of the bumble-bee the maximal activities of fructose diphosphatase are very high and they are similar to the maximal activities of phosphofructokinase (see Table 2; Newsholme et al., 1972; Newsholme & Crabtree, 1973). Furthermore, unlike fructose diphosphatase from every other muscle investigated, the enzyme from bumble-bee flight muscle is not inhibited by AMP. This lack of AMP inhibition may enable both fructose diphosphatase and phosphofructokinase to be simultaneously maximally (or near-maximally) catalytically active in order to generate heat (see Newsholme et al., 1972). It is known that bumble-bees, in common with many other insects, have to maintain a thoracic temperature of about 30°C for flight (see Krogh & Zeuthen, 1941; Heinrich, 1972a,b). However, bumble-bees are unusual in that they forage for food under inclement weather conditions. Heinrich (1972a,b) has shown that the thoracic temperature of the bee can be maintained well above the ambient temperature when the insect is collecting from a flower (the ambient temperature could be as low as 10°C yet the thoracic temperature was near 30°C). Further, observation of the feeding behaviour of bumble-bees shows that they collect

Table 2. *Activities of phosphofructokinase, fructose diphosphatase and glycerol kinase in flight muscle of some bees*

Some of the activities are taken from Newsholme *et al.* (1972) and Newsholme & Crabtree (1973). The values presented are the means of at least three measurements on separate insects in which the variation was not greater than 20% (see above references for ranges of activities).

| Bee | Species | Caste | Enzyme activities (μmol/min per g of flight muscle at 25°C) | | |
			Phosphofructo-kinase	Fructose diphosphatase	Glycerol kinase
Bumble-bee	*Bombus pratorum*	Worker	56	73	2.5
		Male	40	77	—
		Queen	27	25	8.0
	Bombus agrorum	Worker	36	38	1.9
		Male	36	51	—
		Queen	23	14	—
	Bombus lapidarius	Worker	80	114	1.5
		Male	68	70	—
		Queen	27	29	—
	Bombus hortorum	Worker	29	22	2.4
		Male	32	27	—
		Queen	24	10	6.8
'Cuckoo'-bee	*Psithyrus vestalis*	Male	16	1.8	—
		Queen	16	2.2	—
	Psithyrus campestris	Male	17	1.3	—
	Psithyrus sylvestris	Male	18	3.2	—
Honey-bee	*Apis mellifera*	Worker	30	<0.05	2.0
Solitary bee	*Colletes cuniculariusi (celticus O'Toole)*		16	3.3	2.9
	Adrena trimmerana		15	0.7	3.1
	Anthophora acervorum		—	0.9	—

pollen and/or nectar from one flower and fly a short distance to the next flower, so that they may visit a large number of flowers before returning to the hive. The bee alights on a flower and collects food for 1 or 2 min before flying (immediately) to the next flower. Under cold windy conditions, the thorax of the bee would cool continuously during this type of food collection, unless there is some form of heat-generating mechanism which operates during the collecting period. If the bee generated heat (after the collection of the food) by mechanical means (e.g. wing whirring) this would delay the insect during the visits to each flower and decrease the efficiency of food collection. Indeed, Heinrich (1972b) has shown that if a bumble-bee visits a flower that consists of multiple inflorescences, so that the bee can walk from one inflorescence to the next, the thoracic temperature of the bee decreases towards the ambient. Under this situation, the bee does not have to maintain the thoracic temperature until it is ready to return to the nest.

There is a species of bee (*Psithyrus* sp.) that resembles the bumble-bee and parasitizes the nests of the latter. The parasitic bee lays eggs in the nests of the bumble-bee and the workers of this colony collect food to nourish the developing parasitic bees (for this reason they are known as 'cuckoo' bumble-bees). Consequently, the 'cuckoo'-bees do not collect food for feeding the colony. Indeed, when they do collect food (presumably for their own requirements) Sladen (1912) has observed that they are not as efficient as bumble-bees. He described the foraging behaviour of *Psithyrus* as follows. 'The movements of the *Psithyrus*, …are lethargic and awkward. When visiting the flowers in search of food she does not travel systematically from blossom to blossom like an industrious humble-bee (sic), but settling upon a bloom she sips lazily sufficient nectar to satisfy her immediate need…Fatigued by the exertion of obtaining food for herself, she is plainly incapable of the sustained effort that would be needed had she to provide for the wants of her young.' It is, therefore, of considerable interest that the flight muscle of the *Psithyrus* species of bumble-bee does not possess high activities of fructose diphosphatase (see Table 2); the activities are approximately 10% of those of phosphofructokinase, which suggests that the substrate cycle is used to increase sensitivity rather than as a specific heat-generating system. Such a lack of heat generation would provide an explanation for the behaviour of the bee as described by Sladen (1912): we suggest that the bee was not fatigued by her exertion, but it gave the impression of fatigue because of the inability to maintain the thoracic temperature for flight during food collection.

There is no detectable fructose diphosphatase activity in the flight muscle of the honey-bee* (Table 2), which indicates that a substrate cycle between fructose 6-phosphate and fructose diphosphate does not occur in this muscle. (It is of course possible that other substrate cycles exist for heat production in this insect.) However, it is well established that honey-bees do not forage for food under inclement weather conditions. Again Sladen (1912) has described quite admirably the behaviour of the honey-bee in contrast with that of the bumble-bee. 'From July 15 to 27 [1910]…the weather was very unfavourable, with strong cold winds, the temperature often failing to exceed 60°F., and there was very little sunshine; yet even on the worst days a fair quantity of honey was gathered [by a domicile of *B. pratorum*] and stored in the cells, although during this period my honey bees not only failed to gather honey, but consumed all previously gathered stores, and had to be fed to avoid starvation.' It is tempting to speculate that at least some of the difference between the behaviour of these two bees is related to the heat production from the substrate cycle between fructose 6-phosphate and fructose diphosphate during collection of pollen and nectar from flowers.

There is now direct evidence in support of the operation of the cycle between fructose 6-phosphate and fructose diphosphate in relation to heat generation in the bumble-bee. The technique involves glucose labelled with ^3H in the 5 position and labelled with ^{14}C in all positions. Details of this work will not be given here since they have been described in detail by Bloxham *et al.* (1973) and M.G. Clark *et al.* (1973*a*), but the principle of the method is described in Appendix 4. The results of the work indicate that there is no substrate cycling in the flight muscle of the bee during flight, but in non-flying bees there is considerable cycling, which is inversely proportional to the ambient temperature. These results support the original hypothesis of Newsholme *et al.* (1972).

* Fructose diphosphatase is present in the flight muscle of a number of solitary bees (see Table 2) which superficially resemble the honey-bee and can easily be confused with it.

The above discussion suggests that substrate cycling between fructose 6-phosphate and fructose diphosphate occurs in bumble-bee flight muscle in order to generate heat. However, the question arises as to whether the cycle provides sufficient heat to maintain the thoracic temperature at about 30°C. Heinrich (1972a) has measured the thoracic temperature of the bumble-bees (by inserting a thermometer into the thorax of foraging bees immediately on capture) and he has measured the rate of cooling of thoraces from dead bees. Assuming that this rate of cooling is the same in live bees, he concludes that a temperature difference between the thorax and the environment of 27°C can be maintained by the production of 0.54 cal/min per thorax, or about 22 cal/min per g of fresh flight muscle (see also Heinrich, 1972b). Assuming a maximum rate of substrate cycling of 100 μmol/min per g (at 35°C) and a value of 18 kcal for the heat released per revolution of the cycle (see Section F), the maximum rate of heat production is 2 cal/min per g. This represents about 10% of the rate of heat production required. It must, therefore, be considered that other mechanisms of heat production exist in the bumble-bee flight muscle. These could be mechanical and/or metabolic. We suggest that other substrate cycles may contribute to the heat production. Some possible additional substrate cycles are described below.

(a) *Possible additional substrate cycles for heat production in the flight muscle of the bumble-bee.* There is now no doubt that the original hypothesis of Newsholme *et al.* (1972), that fructose diphosphatase in the bumble-bee is involved in a heat-generating substrate cycle, is well established by the work of M. G. Clark *et al.* (1973a) and Bloxham *et al.* (1973). It is suggested that other such metabolic cycles may exist in this animal to supplement the above cycle in heat production. The following is a list of possible cycles. (i) The activities of glycerol kinase in the bumble-bee suggest that a cycle between glycerol and glycerol phosphate, catalysed by glycerol kinase and glycerol phosphate phosphatase, may exist in the flight muscle. Some activities of glycerol kinase are reported in Table 2. Since the activities of glycerol kinase are lower than those of fructose diphosphatase, the rates of heat generation would be less. (ii) The activities of monoglyceride lipase and glycerol kinase in the flight muscle of the bumble-bee (Crabtree & Newsholme, 1972) suggest that a cycle between monoglyceride and fatty acid plus

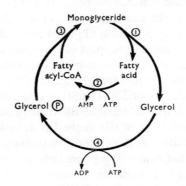

Fig. 8. *Monoglyceride fatty acid cycle*

The reactions are as follows: ① monoglyceride lipase; ② fatty acyl-CoA synthetase; ③ glycerol phosphate acyltransferase; ④ glycerol kinase.

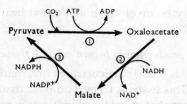

Fig. 9. *Pyruvate–malate cycle in insect flight muscle*

The reactions are as follows: ① pyruvate carboxylase; ② NAD$^+$-linked malate dehydrogenase; ③ NADP$^+$-linked malate dehydrogenase.

glycerol may exist in this tissue (Fig. 8). The enzymes of this cycle would be mono-glyceride lipase, fatty acyl-CoA synthetase, glycerol kinase and glycerol phosphate acyltransferase. Some activities of monoglyceride lipase in the flight muscle of the bumble-bee are given by Crabtree & Newsholme (1972): the highest activity is 0.5 μmol/min per g. Since the conversion of fatty acid into fatty acyl-CoA requires the hydrolysis of ATP to AMP, the total number of ATP molecules hydrolysed in this cycle is three. Nonetheless, since the activities of monoglyceride lipase are rather low the rate of heat generation from this cycle is probably low. (iii) The activities of pyruvate carboxylase and NADP$^+$-linked malate dehydrogenase suggest that a cycle between pyruvate and malate, catalysed by pyruvate carboxy-lase, NAD$^+$-linked and NADP$^+$-linked malate dehydrogenase, may exist in the flight muscle (Fig. 9). Some activities of pyruvate carboxylase and NADP$^+$-linked malate dehydrogenase are given in Table 3. If this cycle is present, its rate of heat generation would be similar to that of the cycle between glycerol and glycerol phosphate, i.e. much less than that of the fructose 6-phosphate/fructose di-phosphate cycle.

If all these cycles do indeed exist in bumble-bee flight muscle and they are switched on during foraging in cold weather, the heat production might approach approximately 50% of that required. In recent years it has been suggested that increases in the activity of the sodium pump may be responsible for some of the heat production in brown adipose tissue (see below). It remains to be seen how

Table 3. *Activities of pyruvate carboxylase and NADP$^+$-linked malate dehydrogenase in the flight muscle of some insects*

Pyruvate carboxylase activities are obtained from Crabtree *et al.* 1972. All activities are means of at least three determinations on different animals. The variation in activities between animals is less than 20%.

Insect	Enzyme activities (μmol/min per g at 25°C)	
	Pyruvate carboxylase	NADP$^+$-linked malate dehydrogenase
Bumble-bee (*Bombus agrorum*)	17.0	6.4
Wasp (*Vespa vulgaris*)	12.0	4.5
Honey-bee (*Apis mellifera*)	4.8	3.2
Yellow underwing moth (*Noctua pronuba*)	2.3	3.9
Locust (*Schistocerca gregaria*)	1.5	2.2
Water bug (*Lethocerus cordofanus*)	2.1	2.5

far this mechanism may be involved in additional heat generation in the flight muscle of the bumble-bee.

2. *Mechanisms of heat production in brown adipose tissue*

Brown adipose tissue is found particularly in hibernating animals and the new-born of some species. This tissue is termed adipose tissue on account of its high lipid content and the fact that in cold-acclimatization some of the white adipose tissue fairly rapidly transforms into brown adipose tissue (i.e. within several days). Nonetheless, the latter tissue is structurally, metabolically and functionally distinct from white adipose tissue. There is now considerable evidence that the role of this tissue is to generate heat. Thus it is important in very young animals, which are born without sufficient insulation to keep them warm in the absence of the mother (e.g. humans, rabbits). It is also important in hibernating animals so that they can resume normal activity after a period of hibernation despite a low ambient temperature [see Smith & Horwitz (1969) for review].

A number of biochemical mechanisms for heat generation in this tissue have been put forward, some of which are indicated below.

(*a*) *Uncoupling of oxidative phosphorylation.* A very obvious mechanism for heat generation is a restriction in oxidative phosphorylation without any restriction in electron transport; this is known as uncoupling. The discovery that a number of compounds can act as uncoupling agents (e.g. dinitrophenol; Loomis & Lipmann, 1948) led to the idea that physiological analogues of such a compound may function to control the degree of coupling in the mitochondria. At times, thyroxine (Hoch & Lipmann, 1954) and long-chain fatty acids (Pressman & Lardy, 1956) have been proposed as candidates for this physiological uncoupler. However, physiological concentrations of thyroxine have no uncoupling effect (see Tata, 1964). Although the concentration of fatty acids that causes uncoupling may be found in some cells, we suggest that there is an intracellular protein that transports long-chain fatty acids within the cell in a similar way to their extra-cellular transport by albumin. In this case, the free concentration of fatty acids in the cell may be too low to cause uncoupling. (Further, it has been suggested that mechanisms may exist to maintain low intracellular concentrations of fatty acids, e.g. the triglyceride–fatty acid cycle in muscle, see Section D.)

The heat generation by brown adipose tissue has been considered to be due largely to uncoupling of mitochondrial oxidative phosphorylation. The experimental basis for this was that mitochondria isolated from the tissue always showed poor P/O ratios in comparison with mitochondria from other tissues (e.g. liver) (Himms-Hagen, 1970). The theory suggested that nervous stimulation releases noradrenaline within the brown adipose tissue and this activates the intracellular triglyceride lipase (via adenylate cyclase and an increase in the intracellular cyclic AMP). The resultant large increase in fatty acid concentration has two effects: the rates of β-oxidation and the tricarboxylic acid cycle are increased, but simultaneously the process of oxidative phosphorylation is uncoupled. Thus there is increased fatty acid oxidation and heat production. However, if mitochondria from brown adipose tissue are prepared under conditions that decrease the fatty acid concentration in the extraction medium, the P/O ratios observed for these mitochondria are not very different from the expected ratios

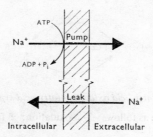

Fig. 10. *Na⁺-translocation cycle*

(see Horwitz & Smith, 1972). Furthermore, other biochemical mechanisms have been put forward to account for some, if not all, of the heat generated in the tissue.

(*b*) *The Na⁺ pump and thermogenesis.* It has been shown that noradrenaline stimulation of thermogenesis in brown adipose tissue is accompanied by a decrease in transmembrane potential, an increase in membrane permeability and an increase in activity of the $(Na^+ + K^+)$-activated ATPase,* which is ouabain-sensitive (Horwitz, 1973; Horwitz & Smith, 1972).

These observations led to the idea that heat generation in brown adipose tissue may be due to an increase in the rate of Na^+ pumping and an increase in permeability of the cell membranes to Na^+. Thus a stimulation of the system that extrudes Na^+ with the hydrolysis of ATP [i.e. the $(Na^+ + K^+)$-activated ATPase] will cause heat production. This heat-generating mechanism is another example of a substrate cycle in which the 'substrates' are translocated rather than transformed: the two reactions of the cycle are the Na^+ 'leak' (entry of Na^+ into the cell along its concentration gradient) and the $(Na^+ + K^+)$-activated ATPase (see Fig. 10).

Recently, it has been suggested that the calorigenic effect of thyroxine may be explained, at least in part, by the effects of this hormone on the Na^+ pump (see Ismail-Beigi & Edelman, 1970). Thus the control of the Na^+-pumping systems may play an important role in heat generation.

The presence of a Na^+ leak and an $(Na^+ + K^+)$-activated ATPase in many tissues other than brown adipose tissue suggests that the 'Na^+-translocation substrate cycle' is a general phenomenon. We consider that the role of this cycle in tissues is to provide a sensitive mechanism for the precise regulation of the intracellular concentration of Na^+. (The sensitivity provided by substrate cycles is discussed in Section C.) It is known that the concentration difference between intracellular and extracellular cations is biologically very important (Williams, 1970). It is further suggested that the 'Na^+-translocation' substrate cycle, which is normally involved in regulation of the intracellular Na^+ concentration, has been extended to provide a specific heat-generating mechanism in brown adipose tissue. (This is similar to the development of the fructose 6-phosphate-fructose diphosphate cycle in the flight muscle of the bumble-bee.)

Although recent experiments have demonstrated that the calorigenic effect of noradrenaline in isolated brown-adipose-tissue cells is produced by effects on

* Abbreviation: ATPase, adenosine triphosphatase.

4

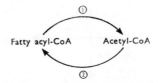

Fig. 11. *Fatty acid oxidation–fatty acid synthesis cycle*

The reactions are as follows: ① β-oxidation; ② fatty acid synthesis.

Na^+ transport, this mechanism can only explain 60% of the calorigenic effect (Horwitz, 1973). Possible mechanisms that might be involved in supplementing the heat-generating mechanism of the Na^+-translocation system are described below.

(*c*) *Other substrate cycles in heat generation in brown adipose tissue.* Other cycles, particularly metabolic cycles, have been suggested to play a role in heat generation in brown adipose tissue. These cycles include the triglyceride–fatty acid cycle, the fatty acid oxidation–fatty acid synthesis cycle and the fructose 6-phosphate–fructose diphosphate cycle. The possible significance of these cycles is discussed below.

(i) The fatty acid oxidation–fatty acid synthesis substrate cycle. There is good evidence that up to 60% of the heat production in brown adipose tissue is produced by the 'Na^+-translocation' cycle (Horwitz, 1973). The ATP that is hydrolysed during the operation of this (and any other) cycle is obtained from the β-oxidation of fatty acids and the subsequent oxidation of acetyl-CoA via the tricarboxylate cycle. However, it has been suggested that some of the acetyl-CoA is reconverted into fatty acyl-CoA by the biosynthetic pathway and this produces another heat-generating cycle (Masoro, 1964) (see Fig. 11). Unfortunately, neither the activities of the enzymes of this biosynthetic system nor the rates of fatty acid synthesis have been systematically studied in brown adipose tissue. Therefore it is not possible to comment on the possible significance of this cycle.

(ii) The triglyceride–fatty acid substrate cycle. This cycle, which is described in Fig. 12, involves the hydrolysis of seven 'energy-rich' phosphate bonds during the re-activation of fatty acids plus glycerol. (This stoicheiometry should be clear when it is realized that fatty acid activation involves the use of two 'energy-rich' phosphate bonds, and that each triglyceride molecule hydrolysed produces three fatty acid molecules and one glycerol molecule.) This cycle has the advantage for heat generation in brown adipose tissue that it requires no provision of components external to the cycle, except ATP. Consequently, the activation of triglyceride lipase alone might be sufficient to stimulate the rate of this cycle and hence the rate of heat production (the ATP is produced from oxidation of a small proportion of the fatty acids obtained from lipolysis).

Attempts have been made to calculate the rate of this cycle from incorporation of [^{14}C]glucose into the glycerol moiety of triglyceride (Knight & Myant, 1970). These results suggest that only a small proportion of heat generated in the brown adipose tissue is obtained from this cycle. Unfortunately, it is possible that the incorporation from exogenous glucose does not give a good indication of the rate

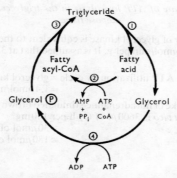

Fig. 12. *Triglyceride–fatty acid cycle in brown adipose tissue*

The reactions are as follows: ① lipolysis; ② fatty acyl-CoA synthetase; ③ esterification; ④ glycerol kinase.

of the cycle (see Smith & Horwitz, 1969). Another approach to this problem is to measure the activities of glycerol kinase. Consequently, the activities of glycerol kinase have been measured in an attempt to assess the significance of this cycle in brown adipose tissue of the newborn rabbit. The activities of glycerol kinase are shown in Table 4 together with the activities in the white adipose tissue of the rat.

The results show that the rate of ATP hydrolysis by the cycle (Table 5) could be about $1.4\,\mu\text{mol/min}$ per g. However, the oxygen-uptake data indicate that noradrenaline-stimulated brown adipose tissue utilizes $180\,\mu\text{mol}$ of ATP/min per g (see Table 5). Thus the triglyceride–fatty acid cycle could contribute only 1% of the total ATP hydrolysis in the brown adipose tissue, if the rate of the cycle was governed by the activity of glycerol kinase. However, it is possible that the glycerol kinase activity is not sufficient to phosphorylate all the glycerol produced from lipolysis and that the rate of glycerol release from this adipose tissue is large. In this case the rates of lipolysis and esterification could be very much greater than calculated above, so that the rate of ATP hydrolysis would be much higher. A problem with this proposal is that glycerol phosphate would have to be supplied

Table 4. *Activities of glycerol kinase, phosphofructokinase and fructose diphosphatase in brown adipose tissue of the neonatal rabbit and white adipose tissue of adult rat*

The activities are presented as means of at least three determinations on different animals. The variation in activities between animals was less than 25%. Activities from the white adipose tissue are taken from Robinson & Newsholme (1967). Other activities are from E. A. Newsholme and T. Williams (unpublished work).

Tissue	Days after birth	Glycerol kinase	Phosphofructo-kinase	Fructose diphosphatase
		Enzyme activities (μmol/min per g fresh weight at 25°C)		
Brown adipose tissue of rabbit	0–1	0.10	0.3	—
	2–3	0.03	0.5	0.02
	4–5	0.10	0.6	0.04
	6–7	0.10	0.7	0.04
White adipose tissue of rat	Adult	0.004	—	—

Table 5. *Calculation of the rate of ATP hydrolysis in the triglyceride–fatty acid cycle in brown adipose tissue*

It is assumed that the activity of glycerol kinase is equivalent to the rate of cycle. The activity of glycerol kinase at 25°C is 0.1 μmol/min per g. It is assumed that at 37°C it would be 0.2 μmol/min per g.

$$\text{Rate of ATP utilization by cycle} = \text{glycerol kinase} \times 7$$
$$= 1.4 \mu\text{mol/min per g}$$

O_2 uptake of noradrenaline-stimulated brown adipose tissue *in vivo* is 3600 μl of O_2/h per 100 mg*
$$\equiv 30 \mu\text{mol of } O_2/\text{min per g}$$
$$\equiv 180 \mu\text{mol of ATP utilized/min per g}$$

* Heim & Hull (1966).

from exogenous glucose (or from endogenous glycogen) and the incorporation from [^{14}C]glucose into the glycerol moiety of triglyceride is low (see above). An alternative suggestion is that the hydrolysis of the triglyceride is only partial, producing fatty acids and monoglyceride. The esterification process would utilize monoglyceride and fatty acyl-CoA so that, although re-activation of fatty acids would be required (see Fig. 8), glycerol would not be involved. This esterification pathway is identical with the monoglyceride shunt pathway in the intestine (see Johnston, 1970). The ATP hydrolysis would be restricted to the conversion of fatty acid into fatty acyl-CoA.

(iii) The fructose 6-phosphate–fructose diphosphate cycle. The activities of fructose diphosphatase and phosphofructokinase in brown adipose tissue of the neonatal rabbit are shown in Table 4. The activities are low, which suggests that the operation of this cycle could not contribute significantly to the heat production of brown adipose tissue.

These activities of phosphofructokinase also provide an indication of the maximum rate of formation of glycerol phosphate for the lipolysis–esterification cycle. They suggest a maximal rate of glycerol phosphate production from glucose of 2.4 μmol/min per g. (The calculation is based on the assumptions that the increase in phosphofructokinase activity for 10°C rise in temperature is twofold and that 1 mol of fructose diphosphate produces 2 mol of glycerol phosphate; thus the activity of phosphofructokinase should be increased fourfold to provide the rate of glycerol phosphate formation.) This would provide sufficient glycerol phosphate for a rate of fatty acid esterification of 7.2 μmol/min per g. This is equivalent to 15 μmol of ATP hydrolysed/min per g, from the triglyceride–fatty acid cycle, which is about 10% of the ATP-turnover rate (calculated from the oxygen uptake, see Table 5). This mechanism could contribute significantly to the heat generated in addition to that of the Na^+-translocation cycle.

F   Control of the Rate of Substrate Cycling

Substrate cycles consist of opposing non-equilibrium reactions so that their operation is accompanied by a continual release of free energy as heat. For example, in the fructose 6-phosphate–fructose diphosphate cycle the heat produced by one revolution of the cycle is the same as that produced by the hydrolysis of one molecule of ATP. Thus the production of heat is an unavoidable consequence of cycling, and cycles have been developed in certain animals for the production of heat (Section E). However, for metabolic regulation the amount of heat released by the cycle may set an upper limit to the amount of sensitivity conferred by cycling.

The sensitivity conferred by a substrate cycle is increased with the magnitude of the cycling rate/flux ratio (see Section C). In theory, any degree of sensitivity can be attained by providing the appropriate value for the ratio but, since the flux is fixed by the metabolic requirement of the pathway, the only variable is the cycling rate. However, it is not possible to increase the rate of cycling indefinitely for at least two reasons. First, the amount of energy that is utilized at a high cycling rate may approach the energy-production capacity of the cell. Secondly, the increased rate of heat production, if it cannot be dissipated, may lead to hyperthermia. Indeed, there is some evidence that an increase in the rate of the substrate cycle between fructose 6-phosphate and fructose diphosphate, in the muscles of susceptible pigs, may be responsible for some of the heat production in a condition known as malignant hyperthermia. This condition is induced by halothane anaesthesia (M. G. Clark et al., 1973b).

In order to estimate the quantity of heat that can be released from substrate cycles, it is necessary to consider a number of factors.

1. *Rate of heat production from substrate cycles*

In most, if not all, cycles heat is generated either directly or indirectly from the hydrolysis of ATP. Thus the rate of heat production from the cycle *per se* can be calculated from the following equation

$$\mathrm{d}H_c/\mathrm{d}t = h_c n C$$

where h_c is the heat released by the hydrolysis of 1 mol of ATP to ADP, which is about 10 kcal/mol; n is the number of molecules of ATP hydrolysed per revolution of the cycle, and C is the rate of cycling in mol/unit time. However, the heat production by the cycle is less than the total heat production as a result of cycling.

Substrate cycling causes ATP hydrolysis which demands rephosphorylation of ADP and concomitant stimulation of metabolic processes to maintain the steady-state concentration of ATP. This will result in further heat production due to the unidirectional nature of both metabolism and oxidative phosphorylation. Although the rate of heat release from the cycle can be calculated from the enthalpy of ATP hydrolysis, it is more difficult to calculate the heat produced from these other metabolic reactions. However, the total heat produced by the hydrolysis of 1 mol of ATP and rephosphorylation of 1 mol of ADP (i.e. that released in the cycle and that released on subsequent stimulation of metabolism) can be calculated from the heat of combustion of various substrates (e.g. carbohydrate, fat) and the number of ATP molecules produced during the oxidation of these fuels (this total heat production is indicated by the symbol h_T). It has been shown that for each molecule of ATP synthesized and hydrolysed 17.4 or 18.1 kcal of heat are released when carbohydrate and fat respectively are utilized as fuels (Krebs,

Table 6. *Heat produced by the fructose 6-phosphate–fructose diphosphate cycle in muscle in the whole animal*

The heat produced is calculated from the equation $dH_T/dt = h_T n C$ where h_T is the total heat produced due to the cycle (see the text), n is the number of ATP molecules hydrolysed to ADP per revolution of the cycle (in this case $n = 1$) and C is the activity of the cycle (in this case $C =$ fructose diphosphatase activity). It is assumed that there is a doubling in the rate of catalysis by the enzyme for an increase in temperature from 25° to 35°C. For the mammals, the fructose diphosphatase activity in the particular muscle is taken as representative of the total musculature, and it is further assumed that the total musculature represents 40% of the total body weight (Munro, 1970). In the chicken, only the contribution from the pectoral muscle is considered, and these muscles contribute 20% to the total body weight (George & Berger, 1966).

Animal	Muscle	$10^6 \times C$ (mol/min)	h_T	$10^6 \times$ (kcal/min per g of muscle)	(kcal/h per kg of animal)	Basal metabolic rate (kcal/h per kg of animal)
Rat	Quadriceps	1.0	18	18.0	0.43	4.1*
Rabbit	Adductor magnus	1.8	18	32.4	0.78	2.5*
Chicken	Pectoral	5.0	18	90.0	1.08	2.0†
Cat	Gastrocnemius	2.6	18	46.8	1.12	2.1*
Man	Undefined leg	0.5	18	9.0	0.22	1.0*

* Fruton & Simmonds (1958).
† Sturkie (1966).

1964).* Consequently, it is possible to calculate the rate of total heat production as a result of substrate cycling. This has been done for the cycle between fructose 6-phosphate and fructose diphosphate and the results are reported in Table 6. The total maximal rate of heat production from this cycle is compared with the basal metabolic rate in several animals (Table 6). The results are very interesting. They indicate that, if this substrate cycle operates at maximal capacity, a considerable amount of heat is generated. The amount is sufficient to contribute about 10–50% of the basal metabolic rate. Since this is the heat produced from only one substrate cycle and since other cycles are known to exist in muscle and other tissues (see Section D), it is unlikely that such cycles can be maximally active for long periods of time. If this were the case excessive heat production would occur, which would result in hyperthermia. Indeed, there is some evidence that this may happen under certain conditions. The administration of halothane to a susceptible strain of pig causes a large increase in heat production, which raises the temperature of the animal to very high and often fatal levels; the condition is known as malignant hyperthermia. M. G. Clark *et al.* (1973b) have demonstrated that, although there was no effect in the normal pig, 30 min after the administration of halothane to the susceptible pig the temperature difference between the arterial and venous blood in the muscle was 2°C and the body temperature was increased by almost 6°C. Under these conditions, the rate of cycling between fructose 6-phosphate and fructose diphosphate was increased.

* The total heat produced by the hydrolysis and rephosphorylation of 1 mol of ATP is calculated by dividing the heat released from the oxidation of 1 mol of fuel by the number of molecules of ATP generated. For example, the oxidation of 1 mol of glucose to CO_2 plus water is accompanied by the release of approximately 710 kcal of heat: in metabolism, 38 mol of ATP are produced so that the total heat released per mol of ATP used in cycling is 710/38, i.e. h_T is approximately 19 kcal/mol. Krebs (1964) has calculated that the value of h_T ranges from 18 kcal/mol for the oxidation of glycerol tristearate to approximately 21 kcal/mol for the oxidation of protein.

2. *The paradox of heat production and sensitivity of control in substrate cycles*

The above discussion indicates that if, in some cycles, maximal rates of cycling occur for any length of time, heat generation may be excessive and result in hyperthermia. However, sensitivity in metabolic control demands a high cycling rate so that the ratio, C/J, is large (see Section C). Thus there appears to be a paradox: cycling rates must be high to provide sensitivity, but if they are too high they may cause over-heating. This problem is particularly relevant to muscle, since this tissue constitutes a large proportion of the body weight, but the link between the ATP-producing metabolic systems and the ATP-requiring systems demands sensitivity in control so that sufficient ATP is produced for the contractile process (see Section D). A hypothesis is proposed in which the rate of substrate cycling may be sufficient to provide sensitivity in metabolic regulation without causing hyperthermia.

(*a*) *The hypothesis of variable rates of substrate cycling in muscle.* It is suggested that in normal resting muscle the rate of substrate cycling between fructose 6-phosphate and fructose diphosphate is very low. For example, it may be only 1–5% of the maximal. This implies that the activities of both phosphofructokinase and fructose diphosphatase are very low under these conditions. However, the activities of both enzymes can be increased under certain conditions in resting muscles, so that the rate of substrate cycling increases dramatically (e.g. to 50% of the maximum), so that the ratio C/J becomes large. It is necessary that this increase in cycling rate occurs for only a short period of time to avoid excessive heat production, but, during this period, the cycle will provide high sensitivity of the net rate of fructose 6-phosphate phosphorylation to changes in concentrations of AMP and other regulators. The increase in cycling could be brought about by hormones or nervous stimuli when the animal anticipates mechanical activity. The extreme situation is exemplified by the stress condition known as 'fight or flight'. Thus adrenaline and/or noradrenaline and sympathetic nervous activity are possible anticipatory stimuli that might increase the rate of cycling. Indeed, adrenaline is known to increase the intracellular [cyclic AMP] which leads to an increase in the activity of phosphorylase and this results in an increased rate of glycogen breakdown (see Section D). The concentration of hexose monophosphate will increase and this will stimulate the activity of phosphofructokinase. If, in addition, the anticipatory hormones lead to a stimulation of the activities of fructose diphosphatase, the rate of this substrate cycle could increase dramatically. Such a transient increase in the rate of cycling would increase the sensitivity of this stage of glycolysis to changes in regulators without the problem of hyperthermia. If the anticipation of exercise is followed by mechanical activity, this would cause changes in the concentrations of metabolic regulators of either of the enzymes of the cycle (e.g. AMP, NH_4^+, P_i, citrate; see Sugden & Newsholme, 1975) that would increase the rate of fructose 6-phosphate phosphorylation sufficiently to ensure that the rate of ATP formation was equal to the rate of ATP utilization. Thus the rate of cycling would be increased only during the anticipatory period to ensure that the regulatory system was sufficiently sensitive to the changes in concentrations of regulators that are produced during the mechanical activity.

If the animal did not need to respond to the anticipatory signals by increased activity, the hormone concentrations would be decreased and the rate of cycling would rapidly return to its original low rate. Hyperthermia would thus be avoided.

At the present time, there is no evidence that the rate of cycling between fructose 6-phosphate and fructose diphosphate can change under different physiological situations. Nonetheless, there is evidence that adrenaline can increase the rate of other substrate cycles. Thus adrenaline increases the rate of the triglyceride–fatty acid cycle in adipose tissue (see Section D) and it probably increases the rate of the glycogen–glucose 1-phosphate cycle in muscle. It should be emphasized that not only does this proposal of transient increases in the rates of substrate cycles explain how increased sensitivity of cycles can occur without over-heating, it also accounts for the well-established calorigenic effect of adrenaline (see Section D and see below).

3. *Calorigenic effect of catecholamines*

A transient increase in the rates of substrate cycling stimulated by the actions of the catecholamines could provide the biochemical basis for the well-established calorigenic effect of these hormones [see Steinberg (1963) and subsequent discussions in the present paper]. Steinberg proposed that the calorigenic effect was due to the mobilization and subsequent oxidation of fatty acids. Thus propanolol inhibited the calorigenic effect and prevented the mobilization of fatty acids. However, Carlson *et al.* (1965) have shown that the calorigenic effect of catecholamines is not always associated with increased mobilization of fatty acids; nicotinic acid inhibits the mobilization of fatty acids, but does not inhibit completely the calorigenic effect of adrenaline. On the other hand, nethalide inhibits both the calorigenic effect and fatty acid mobilization. It is of some importance, therefore, that nethalide, unlike nicotinic acid, inhibits the effects of catecholamines on glycogenolysis in muscle (Carlson *et al.*, 1965). This work is consistent with at least part of the calorigenic effect of the catecholamines being the result of increasing the rates of substrate cycling between glycogen and glucose 1-phosphate and fructose 6-phosphate and fructose diphosphate in muscle.

4. *Substrate cycles and weight control*

If the rates of substrate cycling can be variable, the question can be posed of whether the variation in rate of cycles can be used for anything other than sensitivity in metabolic regulation. Can this variability be used to modulate the rate of conversion of energy, derived from oxidation of foodstuffs, into heat? In this case, the heat release would not be used primarily for temperature regulation, but it would be adjusted for metabolic regulation (see above) and for regulation of the body weight.

It is well established that the body weight of an animal can be maintained remarkably constant despite a large variation in the food intake. Neuman reported in 1902 that his body weight remained constant during a 725-day experiment performed on himself, despite a large variation in his intake of food and beer (quoted by Sims *et al.*, 1973). He coined the term 'Luxuskonsumption' for

Table 7. *Estimates of efficiency with which energy in food is used in different animals*

The data are from Blaxter (1970). Below maintenance diet does not provide sufficient energy to maintain the normal weight of the animal.

Nutrient	Species	Percentage efficiency of utilization of food	
		Below-maintenance diet	Above-maintenance diet
Sucrose	Man	94.2	—
Glucose	Sheep	93.4	54.5
Starch	Domestic fowl	97.2	77.5
Meat	Man	77.4	—
Casein	Sheep	80.6	50.2
Triglyceride	Man	97	—
Aractis oil	Rat	—	83

this ability. Similarly, Gulick (1922) described results of his experiments over 347 days with a subject of the 'difficult fattening' type. His conclusions are summarized as follows. (i) The subject owed his resistance against fattening to an extravagant calorie requirement which persisted at all times despite a moderate daily round of activities. (ii) This extravagance increased during the course of an excessive carbohydrate diet, and it stayed above the initial level after return to normal feeding. (iii) The basal metabolic rate was not involved; it remained strictly normal.

This effect has been studied in animals by Blaxter (1970). He has shown that animals on a diet containing an excess of calories utilize this energy less efficiently than animals maintained on a below-maintenance calorie diet (see Table 7).

More recent experiments have been done on human subjects. Sims *et al.* (1973) summarized their findings from experiments with normal volunteers eating increased amounts of mixed diets as follows. 'Normal subjects fattened by eating a mixed diet require more calories in relation to their body surface area for maintenance of the obese state than they require when at natural weight.' These observations lead us to propose that this ability to maintain the body weight at some 'fixed' level despite variations in caloric intake is due to the effects of variable rates of substrate cycling. By some unknown means, the rates of substrate cycles are depressed when the calorie intake is below maintenance level and they are increased when the intake is above maintenance level. Since radiochemical techniques are now available for the measurement of the rates of substrate cycling (see Katz & Rognstad, 1975) this hypothesis can be readily tested.

The above hypothesis, if valid, has considerable potential in understanding the basic metabolic lesion in obesity. In these patients, the normal control mechanism for maintenance of the natural body weight is impaired. Even if the patients over-eat, which many quite vehemently claim they do not, they do not appear to possess the mechanism for utilization of this excess of calories. Hence any intake of food above the maintenance level causes an increase in the formation of the major storage material triglyceride in the adipose tissue. Passmore *et al.* (1963) studied the weight gain in two obese young women and two thin men when both groups had an excess of caloric intake. The rate of weight gain at any level of calorie intake was considerably greater for Pat and Betty in comparison with that for Sam and Michael. It is suggested that obese patients suffer from decreased rates of substrate cycling or a metabolic inability to relate the rate of cycling to the quantity of calories consumed, so that the normal variable control of the cycling processes is inoperative in obese patients. This hypothesis, as described above, is available for testing, since rates of substrate cycling can now be measured *in vivo* (see Katz & Rognstad, 1975). If the hypothesis proves to be correct, it would open up a new approach to chemotherapy for the obese patient.

G   Substrate Cycles and the Specific Dynamic Action of Food

The basal metabolic rate of an animal is increased after the ingestion of food, and the increase is marked if the animal has previously been starved. The increase in rate begins about 1 h after the meal and lasts for several hours, after which the rate returns to its original value. The effect has been observed after ingestion of all three major categories of food, i.e. protein, fat or carbohydrate. The effect is known as the specific dynamic action of food (Rubner, 1902; Blaxter, 1962; Mitchell, 1964; Krebs, 1964), and it is usually expressed as the amount of heat released during the period of elevated metabolism (compared with the starvation basal state) as a percentage of the calorific value of the food (see Mitchell, 1964). The magnitude of the effect varies according to the animal, its previous dietary history and the food. An indication of magnitude is given by the following approximate values: carbohydrate, 5–20%; fat, 5–10%; protein, 5–40%. Thus the amount of heat released in this way is not insignificant.

The problem has been studied at various levels of intensity during this century, but no satisfactory biochemical explanation has been put forward for the specific dynamic action of all foods. A hypothesis is advanced that this effect of food can be explained by an increase in the rate of substrate cycling during metabolism of the digested products of the food. In order to present this hypothesis, the characteristics of the specific dynamic action will be described briefly, the previous theories will be adumbrated and a simple hypothesis of substrate cycling and specific dynamic action will be described. In this way, it can be seen how substrate cycling accounts for all the known characteristics of the specific dynamic action of food.

1. *Characteristics of the specific dynamic action of food*

Since many of these characteristics are discussed in detail by Mitchell (1962; 1964) they are given in outline only. (*a*) The increase in heat production varies with the type of food and the animal. (*b*) The extent and duration of the heat production is proportional to the amount of food ingested. (*c*) The increase can be observed after injection of amino acids, so that it is not solely due to the digestive processes. (*d*) The increase is not dependent on oxidation of the food. Thus administration of glucose to dogs which had been starved for 3 weeks produced a pronounced specific dynamic action, although most of the glucose was excreted in the urine (Dann & Chambers, 1930). (*e*) The increase in heat production from carbohydrate and fat is abolished if exercise is taken after the meal. There is no decrease in that from protein. (*f*) Removal of the liver from dogs abolishes the specific dynamic action of protein but not that of carbohydrate or fat. (*g*) The specific dynamic action is either zero or positive, it is never negative. Thus changing from one food to another does not decrease the basal metabolic rate.

These properties are consistent with the hypothesis that specific dynamic action is due to substrate cycling (see below).

2. *Theories of specific dynamic action*

The idea that this effect of food might result from increased metabolism due to an increased concentration of metabolites in the blood was proposed as early as 1881 by Voit [cited by Mitchell (1964)]. The hypothesis, which could be described as that of 'plethora', was supported by experimental evidence of Lusk and his collaborators (see Mitchell, 1964). However, the increased metabolism that was required by the hypothesis was assumed to represent oxidation of the food and some of the properties of the specific dynamic action conflict with this (see above). Consequently, the 'plethora' hypothesis was abandoned in spite of its ability 'to explain fairly completely the observed facts concerning the specific dynamic action of sugars and fats and the factors affecting it' (Mitchell, 1964).

In 1964 Krebs proposed that the specific dynamic action of protein could be explained by the energy requirements of urea synthesis and wastage of some of the energy in the amino acid-degradation pathways (i.e. there are reactions in the degradative pathways that are not biological dehydrogenations or do not conserve energy in the formation of ATP). These proposals may explain why the heat production from protein is usually greater than that from carbohydrate and fat, but the proposal does not explain the heat production from the latter foods and why the production of heat should be so variable. Moreover, this theory predicts a decreased rate of heat production when fat or carbohydrate oxidation replaces that of protein: such a negative specific dynamic action has never been observed (see above).

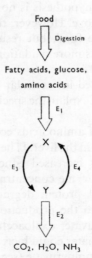

Fig. 13. *A substrate cycle in parallel with oxidation of metabolites*

The reactions catalysed by enzyme systems E_1 and E_2 represent the normal catabolic processes. E_3 and E_4 represent reactions that constitute one (or more) substrate cycles, which are stimulated by increases in the concentrations of X and Y.

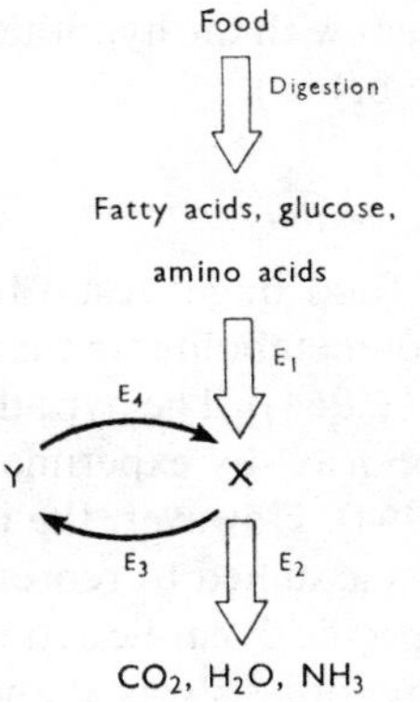

Food

Digestion

Fatty acids, glucose,

amino acids

E_1

E_4

Y   X

E_3 E_2

CO_2, H_2O, NH_3

Fig. 14. *A substrate cycle independent of oxidation of metabolites*

The reactions catalysed by E_1 and E_2 represent the normal catabolic processes. E_3 and E_4 represent reactions that constitute one (or more) substrate cycles which can be independent of oxidation. An increase in the rate of E_1 (due to increased concentration of fatty acids, amino acids or glucose) could increase X which would increase the rate of cycling but may not increase the rate of reaction E_2.

3. *Substrate cycling as a mechanism to explain specific dynamic action*

The explanation for the heat production after ingestion of food, involving the concept of substrate cycling, is very straightforward. It is suggested that the metabolism of the digested products of food increases the concentrations of metabolic intermediates in the major metabolic pathways. These increased concentrations of metabolites increase the rate of substrate cycling and this causes the production of heat. Indeed, the hypothesis is no more than an extension of the theory of 'plethora' discussed above. However, the increased metabolism is not directly due to oxidation of the food, but results from the increased rate of substrate cycling, which demands more oxidation of food to provide the energy for the cycles (see Figs. 13 and 14).

The possible effects of increased intracellular concentrations of amino acids, glucose and fatty acids that might explain the specific dynamic action are indicated as follows.

(i) Increased concentrations of amino acids could increase the rate of cycling between amino acids and protein in the liver. [The activity of the protein-synthesis system might be stimulated by increased concentration of amino acids, see Munro (1970). If protein and enzyme concentrations remain constant this can only be explained by increased protein degradation, i.e. increased cycling.] Amino acid metabolism leads to the production of gluconeogenic precursors (e.g. pyruvate, phosphoenolpyruvate, oxaloacetate) so that an increase in the concentrations of these intermediates could increase the rate of substrate cycling between pyruvate and phosphoenolpyruvate (see Fig. 2).

(ii) An increase in the intracellular glucose concentration (due to increased blood glucose concentration) would be expected to increase cycling between glucose and glucose 6-phosphate, and between glucose 1-phosphate and glycogen in muscle or liver (see Section D). Further, an increase in hexose monophosphate concentrations could increase the rate of cycling between fructose 6-phosphate and fructose diphosphate.

(iii) An increase in the blood content of triglyceride and fatty acids would be expected to increase the rate of cycling between fatty acids and triglyceride in both muscle and adipose tissue (see Section D). The rate of the 'triglyceride' cycle between liver and adipose tissue (Section D) may also be increased.

The properties of the specific dynamic action of foods (see above) are entirely consistent with the hypothesis that the biochemical basis of this phenomenon is increased substrate cycling due to increased concentrations of metabolites. Thus the variation in the heat production with the food and between animals is explained by the variation in capacities of the different cycles and the probable variation between animals. The greater the amount of ingested food, the higher the concentrations of metabolites, the greater the rate of cycling and heat production (see above). Since mechanical activity would be expected to increase metabolism of both fat and carbohydrate so that the cycling rates might already be saturated, increases in the intracellular or extracellular concentrations of glucose or fat would have no further effect on cycling rates. Since cycles involving glucose or fat occur in muscle and adipose tissue, removal of the liver would be expected to have little effect on the specific action of these foods. In particular, the hypothesis predicts that actual oxidation of glucose is not required to produce a specific dynamic effect. Thus an increase in glucose concentration could increase the rate of cycling between glucose and glucose 6-phosphate without any change in the rate of utilization of the glucose. Obviously, some other substrate must be used to provide the energy for cycling but this could be provided by oxidation of fatty acids. Since this specific dynamic action of glucose, without concomitant oxidation, was observed in dogs starved for 3 weeks (Dann & Chambers, 1930), it is likely that most of the energy is produced from fat oxidation. It is interesting that the observation, which was totally inconsistent with the theory of 'plethora' of Voit, is entirely consistent with the hypothesis that substrate cycles are responsible for the specific dynamic action of foods.

H Alcoholic Hypothermia

There is a considerable amount of evidence linking alcoholic intoxication and hypothermia. Reincke, in 1925, in 'Observations on Body Temperature in Drunkards' reported 17 cases of hypothermia associated with acute alcoholic intoxication (quoted by Weyman et al., 1974). A patient who survived the lowest recorded temperature (18°C) was intoxicated (Laufman, 1951). In 1963, Andersen et al. commented 'that grave alcoholic intoxication combined with exposure to a cold environment may be lethal is proved by the numerous case histories of drunken persons who have fallen asleep in the cold and frozen to death'. However, alcoholic intoxication does not appear to produce hypothermia directly, but it lowers the ability of the subject to produce more heat in response to a very low ambient temperature. Anderson et al. (1963) suggest that moderate doses of alcohol had no deleterious effect on the heat balance during prolonged mild cold exposure. However, under conditions of severe cold stress, one of the experimental subjects had insufficient metabolic compensation to the cold after alcohol.

Table 8. *Effect of alcohol and cold exposure after exhaustive exercise on blood glucose concentrations and body temperature*

The data are taken from Haight & Keatinge (1973). These are the mean values from seven subjects. Exercise was running on a treadmill or pedalling a bicycle ergometer until exhaustion. All subjects were subjected to cold exposure (see the text). Glucose was administered after the exercise.

| | No alcohol | | Alcohol | |
Condition	Rectal temperature (°C)	Blood glucose (mM)	Rectal temperature (°C)	Blood glucose (mM)
Resting	36.7	3.5	36.8	2.8
Exercise	36.6	3.1	34.5	1.8
Exercise and glucose	36.7	6.2	36.9	5.0

Haight & Keatinge (1973) investigated the effect of cold exposure on body temperature of male volunteers aged 19–27, who were starved, exercised to exhaustion and placed in a cold-room (14.5°C) in an airflow of 15 km/h wearing only shorts. Ethanol was given orally after the exercise and it was observed that the rectal temperature was lowered after exposure to the cold. The blood glucose concentration was also measured and this was lowered after the ethanol and cold exposure (Table 8). If glucose was administered orally with the ethanol, neither the blood glucose concentration nor the rectal temperature was lowered on exposure to cold. Consequently, these workers concluded that hypoglycaemia was the cause of the hypothermia. They suggested that lowered blood glucose concentrations caused an impaired function of the hypothalamic temperature-regulating centre, leading to poor control of the body temperature. In a completely different situation, a relationship between hypoglycaemia and hypothermia was noted in patients in hospital by Kedes & Field (1964). They suggested that hypothermia might be indicative of hypoglycaemia.

The metabolic effect of alcohol, which may explain the hypoglycaemia observed by Haight & Keatinge (1973) and other workers, is an inhibition of gluconeogenesis in the liver (see Krebs, 1968). The biochemical details of the inhibition are not fully understood, but an increase in the reduced state of the $NAD^+/NADH$ redox couple may be partially responsible. The contents of some of the metabolic intermediates of gluconeogenesis in the liver are markedly decreased after alcohol administration (Williamson *et al.*, 1969). In particular, the contents of glucose, glucose 6-phosphate, fructose 6-phosphate, pyruvate and phosphoenolpyruvate are decreased. We suggest that decreases in these concentrations could decrease the rate of four substrate cycles in liver: glucose and glucose 6-phosphate; glycogen and glucose 1-phosphate; fructose 6-phosphate and fructose diphosphate; pyruvate and phosphoenolpyruvate (see Fig. 2). The hypoglycaemia may also lower the concentrations of hexose monophosphate in muscle and thus decrease the rate of cycling between glucose 1-phosphate and glycogen and between fructose 6-phosphate and fructose diphosphate. Consequently, the relationship between alcoholic intoxication, hypothermia and hypoglycaemia may be explained by depressed rates of substrate cycling in liver and muscle. The inhibition of gluconeogenesis by alcohol would be the primary event leading to both decreased rates of cycling and hypoglycaemia. The inability to increase the rates of these cycles in the intoxicated person could account for the observed failure to respond by producing more heat in severe cold exposure.

I Accidental Hypothermia in the Elderly

Hypothermia is diagnosed when the rectal temperature is below 95°F, and the term accidental hypothermia applies to conditions that are not induced under medical supervision. In 1958 in the *Lancet*, Emslie-Smith pointed out that the condition of accidental hypothermia 'is very much commoner than is supposed'. This condition was diagnosed by Emslie-Smith in nine elderly patients. Duguid *et al.* (1961) emphasized the problem as follows: 'in recent years attention has been drawn to a form of hypothermia mainly in debilitated elderly persons in no more hostile an environment than their own home'. A series of 23 elderly patients were reported to be suffering from accidental hypothermia, which had developed indoors. Only 7 of these patients recovered.

In 1962, Prescott *et al.* reported that over a 4-month winter period, 2.5% of the acute medical admissions to a London hospital were accidental hypothermia. The rectal temperature ranged from 75–95°F and the age ranged from 44 to 82 years. It has been calculated that during the 3 months of a mild winter in 1968, 9000 patients may have been admitted to hospitals in Great Britain suffering from hypothermia (Agate, 1971).

The possibility is put forward that hypothermia in the elderly may be due to decreased rates of substrate cycling in tissues such as the liver and muscle. If rates of cycling are decreased, the heat-generating capacity of these tissues would be impaired. In this case, elderly persons would not be able to respond to even mild cold conditions by increasing heat generation. Whether this possibility has any validity must await investigations into the variation of the rates of substrate cycling with age.

We thank Dr. A. R. Leech who prepared the Figures and Mrs. Thelma Williams who performed the enzyme activity assays reported in Tables 2, 3 and 4.

References

Agate, J. (1971) in *Clinical Geriatrics* (Rossman, I., ed.), pp. 469–471, J. B. Lippincott and Co., Philadelphia and Toronto
Andersen, K. L., Hellstrom, B. & Lorentzen, F. V. (1963) *J. Appl. Physiol.* 18, 975–982
Blaxter, K. L. (1962) *The Energy Metabolism of Ruminants*, Hutchinson, London
Blaxter, K. L. (1970) *Fed. Proc. Fed. Am. Soc. Exp. Biol.* 30, 1436–1443
Bloxham, D. P., Clark, M. G., Holland, P. C. & Lardy, H. A. (1973) *Biochem. J.* 134, 581–587
Bücher, Th. & Rüssmann, W. (1964) *Angew. Chem. Int. Ed. Engl.* 3, 426–439
Carlson, L. A. (1969) *Biochem. J.* 114, 49 P
Carlson, L. A., Boberg, J. & Hogstedt, B. (1965) in *Handbook of Physiology: Section 5: Adipose Tissue* (Renold, A. E. & Cahill, G. F., eds.), pp. 625–644, American Physiological Society, Washington, DC
Clark, D. G., Rognstad, R. & Katz, J. (1973) *Biochem. Biophys. Res. Commun.* 54, 1141–1148
Clark, M. G., Bloxham, D. P., Holland, P. C. & Lardy, H. A. (1973a) *Biochem. J.* 134, 589–597
Clark, M. G., Williams, C. H., Pfeifer, W. F., Bloxham, D. P., Taylor, C. A. & Lardy, H. A. (1973b) *Nature (London)* 245, 99–101
Clark, M. G., Kneer, N. M., Bosch, A. L. & Lardy, H. A. (1974) *J. Biol. Chem.* 249, 5695–5703
Crabtree, B. & Newsholme, E. A. (1972) *Biochem. J.* 130, 697–705

Crabtree, B. & Newsholme, E. A. (1975) in *Insect Muscle* (Usherwood, P. N. R., ed.), pp. 405–500, Academic Press, London and New York
Crabtree, B., Higgins, S. J. & Newsholme, E. A. (1972) *Biochem. J.* **130**, 391–396
Dann, M. & Chambers, W. H. (1930) *J. Biol. Chem.* **89**, 675–688
Denton, R. M. & Randle, P. J. (1967) *Biochem. J.* **104**, 416–422
Denton, R. M., Yorke, R. E. & Randle, P. J. (1966) *Biochem. J.* **100**, 407–419
Duguid, H., Simpson, R. G. & Stowers, J. M. (1961) *Lancet* **ii**, 1213–1219
Emslie-Smith, D. (1958) *Lancet* **ii**, 492–495
Fischer, E. H., Pocker, A. & Saari, J. C. (1970) *Essays Biochem.* **6**, 23–68
Fruton, J. S. & Simmonds, S. (1958) *General Biochemistry*, 2nd edn., p. 936, Wiley, New York and London
George, J. C. & Berger, A. J. (1966) *Avian Myology*, Academic Press, New York and London
Gordon, R. S. & Cherkes, A. (1956) *J. Clin. Invest.* **35**, 206–212
Gulick, A. A. (1922) *Am. J. Physiol.* **60**, 371–395
Haight, J. S. J. & Keatinge, W. R. (1973) *J. Physiol. (London)* **229**, 87–97
Havel, R. J. & Fredrickson, D. S. (1956) *J. Clin. Invest.* **35**, 1025–1032
Heim, T. & Hull, D. (1966) *J. Physiol. (London)* **186**, 42–55
Heinrich, B. (1972a) *Science* **175**, 185–187
Heinrich, B. (1972b) *J. Comp. Physiol.* **77**, 49–64
Himms-Hagen, J. (1970) *Adv. Enzyme Regul.* **8**, 131–151
Hoch, F. L. & Lipmann, F. (1954) *Proc. Natl. Acad. Sci. U.S.A.* **40**, 909–921
Horwitz, B. A. (1973) *Am. J. Physiol.* **224**, 352–355
Horwitz, B. A. & Smith, R. E. (1972) in *International Symposium on Environmental Physiology and Biophysics* (Smith, R. E., Hannon, J. P., Shields, J. L. & Horwitz, B. A., eds.), pp. 134–140, Federation of American Society for Experimental Biology, Washington, DC
Hue, L. & Hers, H. G. (1974) *Biochem. Biophys. Res. Commun.* **58**, 540–548
Ismail-Beigi, F. & Edelman, I. S. (1970) *Proc. Natl. Acad. Sci. U.S.A.* **67**, 1071–1078
Johnston, J. M. (1970) *Compr. Biochem.* **18**, 1–18
Katz, J. & Rognstad, R. (1975) *Curr. Top. Cell. Regul.* in the press
Kedes, L. H. & Field, J. B. (1964) *N. Engl. J. Med.* **271**, 785–787
Knight, B. L. & Myant, N. B. (1970) *Biochem. J.* **119**, 103–111
Krebs, H. A. (1964) in *Mammalian Protein Metabolism* (Munro, H. N. & Allison, J. B., eds.), vol. 1, pp. 125–176, Academic Press, New York and London
Krebs, H. A. (1968) *Adv. Enzyme Regul.* **6**, 467–480
Krebs, H. A. & Veech, R. L. (1970) in *Pyridine Nucleotide Dependent Dehydrogenases* (Sund, H., ed.), pp. 413–438, Springer-Verlag, Berlin, Göttingen and Heidelberg
Krogh, A. & Zeuthen, E. (1941) *J. Exp. Biol.* **18**, 1–10
Laufman, H. (1951) *J. Am. Med. Assoc.* **147**, 1201–1212
Loomis, W. F. & Lipmann, F. (1948) *J. Biol. Chem.* **173**, 807–808
Masoro, E. J. (1964) *Fed. Proc. Fed. Am. Soc. Exp. Biol.* **22**, 868–873
Masoro, E. J., Rowell, L. B., McDonald, R. M. & Steiert, B. (1966) *J. Biol. Chem.* **241**, 2626–2634
Mitchell, H. H. (1962) *Comparative Nutrition of Man and Domestic Animals*, vol. 1, pp. 3–90, Academic Press, New York and London
Mitchell, H. H. (1964) *Comparative Nutrition of Man and Domestic Animals*, vol. 2, pp. 471–566, Academic Press, New York and London
Munro, H. N. (1970) in *Mammalian Protein Metabolism* (Munro, H. N., ed.), vol. 4, pp. 299–387, Academic Press, New York and London
Newsholme, E. A. (1972) *Cardiology* **56**, 22–34
Newsholme, E. A. (1976) *Prog. Liver Dis.* **5**, in the press
Newsholme, E. A. & Crabtree, B. (1970) *FEBS Lett.* **6**, 195–198
Newsholme, E. A. & Crabtree, B. (1973) in *Rate Control of Biological Processes* (Davies, D. D., ed.), pp. 429–460, Cambridge University Press, London
Newsholme, E. A. & Gevers, W. (1967) *Vitam. Horm. (N.Y.)* **25**, 1–87
Newsholme, E. A. & Start, C. (1972) in *Handbook of Physiology: Section 7: Endocrinology I* (Steiner, D. F. & Freinkel, N., eds.), pp. 369–383, American Physiological Society, Washington, DC
Newsholme, E. A. & Start, C. (1973) *Regulation in Metabolism*, Wiley and Sons, London, Sydney and Toronto
Newsholme, E. A. & Underwood, A. H. (1966) *Biochem. J.* **99**, 24 c–26 c
Newsholme, E. A., Crabtree, B., Higgins, S. J., Thornton, S. D. & Start, C. (1972) *Biochem. J.* **128**, 89–97
Passmore, R., Strong, J. A., Swindells, Y. E. & El Din, N. (1963) *Br. J. Nutr.* **17**, 373–383
Prescott, L. F., Peard, M. C. & Wallace, I. R. (1962) *Br. Med. J.* **2**, 1367–1370

Pressman, B. C. & Lardy, H. A. (1956) *Biochim. Biophys. Acta* **21**, 458–466
Randle, P. J., Garland, P. B., Hales, C. N., Newsholme, R. A., Denton, R. M. & Pogson, C. I. (1966) *Recent Prog. Horm. Res.* **22**, 1–48
Robinson, D. S. (1970) *Compr. Biochem.* **18**, 51–116
Robinson, D. S. (1976) *Proc. Nutr. Soc.* in the press
Robinson, J. & Newsholme, E. A. (1967) *Biochem. J.* **104**, 2c–4c
Rolleston, F. S. (1972) *Curr. Top. Cell. Regul.* **5**, 47–75
Rubner, M. (1902) *Die Gesetze des Energieverbrauchs bei der Ernährung*, Deuticke, Leipzig and Vienna
Rydström, J., Teizeira da Cruz, A. & Ernster, L. (1970) *Biochem. J.* **116**, 12p–13p
Schoenheimer, R. & Rittenberg, D. (1936) *J. Biol. Chem.* **114**, 381–396
Scow, R. O. & Chernick, S. S. (1970) *Comp. Biochem. Physiol. B* **18**, 19–49
Sims, E. A. H., Danforth, E., Horton, E. S., Bray, G. A., Glennon, J. A. & Salans, L. B. (1973) *Recent Prog. Horm. Res.* **29**, 457–496
Sladen, F. W. L. (1912) *The Humble Bee: Its Life History and How to Domesticate It*, Macmillan, London
Smith, R. E. & Horwitz, B. A. (1969) *Physiol. Rev.* **49**, 330–425
Start, C. & Newsholme, E. A. (1968) *Biochem. J.* **107**, 411–415
Start, C. & Newsholme, E. A. (1970) *FEBS Lett.* **6**, 171–173
Steinberg, D. (1963) in *Control of Lipid Metabolism* (Grant, J. K., ed.), pp. 111–138, Academic Press, London and New York
Steiner, D. F. & King, J. (1964) *J. Biol. Chem.* **239**, 1291–1298
Sturkie, P. D. (1966) *Avian Physiology*, 2nd edn., p. 245, Cornell University Press, New York
Sugden, P. H. & Newsholme, E. A. (1975) *Biochem. J.* **150**, 113–122
Tata, J. R. (1964) in *Action of Hormones on Molecular Processes* (Litwack, G. & Kritchevsky, D., eds.), pp. 58–131, Wiley, New York
Underwood, A. H. & Newsholme, E. A. (1967) *Biochem. J.* **104**, 300–305
Weyman, A. E., Greenbaum, D. M. & Grace, W. J. (1974) *Am. J. Med.* **56**, 13–21
Williams, R. J. P. (1970) *Q. Rev. Chem. Soc.* **24**, 331–365
Williamson, J. R., Scholz, R., Browning, E. T., Thurman, R. G. & Fukami, M. H. (1969) *J. Biol. Chem.* **244**, 5044–5054

APPENDIX 1

Relationship between Reversibility (R) and Mass–Action Ratio (Γ)

$$R = \frac{V_f}{V_f - V_r} \text{ and } \frac{K}{\Gamma} = \frac{V_f}{V_r} \text{ (see Section B1c and Appendix 2)}$$

$$\therefore \frac{1}{R} = \frac{V_f - V_r}{V_r} = 1 - \frac{V_r}{V_f} = 1 - \frac{1}{K/\Gamma}$$

$$\therefore \frac{R}{1} = \frac{(K/\Gamma) - 1}{K/\Gamma}$$

$$\therefore R = \frac{(K/\Gamma)}{(K/\Gamma) - 1}$$

A plot of R against K/Γ (see Table 9) indicates that R is greatest when (K/Γ) is between 1 and approx. 5. [When (K/Γ) is unity, R is infinite.] Thus the sensitivity of a reaction to a reactant is largest in the range $K/\Gamma = 1$–5, so that a value of

Table 9. *Relationship between K/Γ and R*

K/Γ	R	Percentage increase in sensitivity (to a substrate) compared with a non-equilibrium reaction {i.e. 100[(R−1)/1]}	
1.01	101	10000	
1.1	11	1000	
1.2	6	500	
1.5	3	200	
2.0	2	100	
4.0	1.3	30	
5.0	1.25	25	(Empirical division between near-equilibrium and
10.0	1.11	11	non-equilibrium reactions)
20.0	1.05	5	
50.0	1.02	2	
100.0	1.01	1	
1000.0	1.001	0.1	

K/Γ equal to 5 may provide a better empirical division between near-equilibrium and non-equilibrium reactions than previous divisions (see Rolleston, 1972).

Reference

Rolleston, F. S. (1972) *Curr. Top. Cell. Regul.* **5**, 47–75

APPENDIX 2

Relationship between the Ratio V_f/V_r and the Ratio K/Γ

Classical thermodynamics relates the free energy change of a reaction (ΔG) to its displacement from equilibrium. Consider the reaction

$$A \rightleftharpoons B$$

The Gibbs free-energy equation states that.

$$\Delta G = \Delta G^0 + RT \ln \frac{[B]}{[A]} \tag{1}$$

where ΔG^0 is the standard free energy change, and [A] and [B] are the concentrations of substrate and product. When the reaction is at equilibrium, ΔG is zero and hence

$$\Delta G^0 = -RT \ln \frac{[B_e]}{[A_e]} = -RT \ln K \tag{2}$$

where $[A_e]$ and $[B_e]$ are the equilibrium concentrations of the reactants and K is the equilibrium constant. Combination of eqns. (1) and (2) gives

$$\Delta G = -RT \ln K + RT \ln \frac{[B]}{[A]} \tag{3}$$

Therefore

$$\Delta G = -RT \ln \frac{[A]}{[B]} K \tag{4}$$

The ratio of the concentrations of products to those of substrates [B]/[A] as measured in a tissue is known as the mass–action ratio and is designated by the symbol Γ (see Bücher & Rüssmann, 1964):

$$\Gamma = \frac{[B]}{[A]} \qquad (5)$$

Eqn. (4) becomes

$$\Delta G = -RT \ln \left(\frac{1}{\Gamma}\right) K$$

i.e.

$$\Delta G = -RT \ln \frac{K}{\Gamma} \qquad (6)$$

The ratio K/Γ is a measure of the displacement of a given reaction from equilibrium (see Bücher & Rüssmann, 1964; Rolleston, 1972): its value is unity when the reaction is at equilibrium and becomes larger the further the reaction is displaced from equilibrium.

If the reaction $A \rightleftharpoons B$ is considered from a kinetic viewpoint, the following equations can be developed

$$\longrightarrow A \underset{V_r}{\overset{V_f}{\rightleftharpoons}} B \longrightarrow$$

Let

$$V_f = k_{+1}[A] \qquad (7)$$

and

$$V_r = k_{-1}[B] \qquad (8)$$

where V_f and V_r are the rates of reaction in the forward and reverse directions respectively; k_{+1} and k_{-1} are the rate constants of the forward and reverse directions respectively. Combining eqns. (7) and (8) gives

$$\frac{V_f}{V_r} = \frac{k_{+1}}{k_{-1}} \frac{[A]}{[B]} \qquad (9)$$

At equilibrium,

$$\frac{k_{+1}}{k_{-1}} = \frac{[B_e]}{[A_e]} = K \qquad (10)$$

Therefore substituting from eqn. (9)

$$\frac{V_f}{V_r} = K \frac{[A]}{[B]} \qquad (11)$$

From eqn. (5)

$$\frac{[A]}{[B]} = \frac{1}{\Gamma}$$

Hence

$$\frac{V_f}{V_r} = K \left(\frac{1}{\Gamma}\right) = \frac{K}{\Gamma} \qquad (12)$$

Since the ratio K/Γ is a measure of the displacement of a reaction from equilibrium it follows that the ratio of the reaction rates in the forward and reverse directions (V_f/V_r) must be an identical measure of the displacement of a reaction from equilibrium.

References

Bücher, Th. & Rüssmann, W. (1964) *Angew. Chem. Int. Ed. Engl.* 3, 426–439
Rolleston, F. S. (1972) *Curr. Top. Cell. Regul.* 5, 47–75

APPENDIX 3

Derivation of a Function to Describe Sensitivity in a Substrate Cycle to Changes in Substrate Concentration for the Reverse Reaction

The cycle is described diagrammatically as follows (see Section C2c):

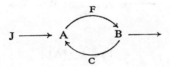

the net flux, $J = F - C$. As before

$$J + \Delta J = F - (C + \Delta C)$$

$$\therefore \Delta J = -\Delta C$$

(The minus sign indicates that an increase in C decreases the net flux.)

$$\frac{\Delta J}{J} = J_{rel.} = \frac{-\Delta C}{J}$$

$$= \frac{-\Delta C}{J} \times \frac{C}{C}$$

$$= \frac{-\Delta C}{C} \times \frac{C}{J}$$

$$= -C_{rel.} \times \frac{C}{J}$$

$$\therefore \frac{J_{rel.}}{C_{rel.}} = -\frac{C}{J}$$

If an effector X interacts linearly with C, so that $C_{rel.} = [X]_{rel.}$ (Section C2c) the sensitivity of the net flux to X (which may be either a mass–action or an allosteric effector of C), $J_{rel.}/[X]_{rel.}$ equals $-C/J$.

APPENDIX 4

Measurement of the Rate of Substrate Cycling between Fructose 6-Phosphate and Fructose Diphosphate

This ingenious method was first described by Bloxham *et al.* (1973) and a similar method has been used by Clark *et al.* (1973). The principle is as follows. When glucose is converted into triose phosphates via glycolysis, the hydrogen atom attached to the fifth carbon atom of glucose exchanges with the hydrogens in water. This is a consequence of the catalytic mechanism of the enzyme triose phosphate isomerase. Since this reaction is near-equilibrium in the glycolytic pathway most if not all of this hydrogen should be exchanged. Thus, if ^3H is attached to the fifth carbon of the glucose molecule it will be lost at the triose phosphate isomerase reaction. If such glucose is administered to an intact tissue, the ^3H will be lost to the intracellular water. Consequently the specific radioactivity of the triose phosphates, relative to ^3H, becomes close to zero. The near-equilibrium nature of the aldolase reaction results in dilution of the specific radioactivity of fructose diphosphate by the unlabelled triose phosphates, so that this may also become close to zero. Consequently, if there is cycling between fructose 6-phosphate and fructose diphosphate, the specific radioactivity of fructose 6-phosphate (and hence the hexose monophosphates) will also be decreased (i.e. the amount of ^3H in position 5 of hexose phosphates will be less than that originally present).

Let it be assumed that the system *in vivo* can be represented as follows [the symbols and derivations used here differ somewhat from those used by Bloxham *et al.* (1973)]:

In this system G denotes the rate of glycolysis (i.e. glycogenolysis is neglected), and F and C denote the rates of the reactions catalysed by phosphofructokinase and fructose diphosphatase respectively; the reaction catalysed by phosphoglucose isomerase is assumed to be sufficiently close to equilibrium that glucose 6-phosphate and fructose 6-phosphate can be considered as a single pool of hexose monophosphates. Let the steady-state specific radioactivities of glucose, hexose phosphate and fructose diphosphate (relative to ^3H in position 5 of glucose) be s_g, s_p and s_f respectively; the specific radioactivity of triose phosphate is assumed to be zero (see above). The rate of labelling of hexose phosphate with ^3H in the steady state is described by the equation

$$G \cdot s_g + C \cdot s_f = F \cdot s_p$$

so that

$$s_p = (G \cdot s_g + C \cdot s_f)/F$$

Expressed as a specific radioactivity relative to that of the administered glucose (s_g), this equation becomes

$$s_p/s_g = G/F + (C/F)(s_f/s_g)$$

Let the ratios (s_p/s_g) and (s_f/s_g) be denoted by the symbols s_1 and s_2 respectively; the above equation may then be transformed, by the relation $G = F - C$, into the equation

$$s_1 = 1 - (C/F)(1 - s_2) \qquad (1)$$

Thus a determination of the values of s_1 and s_2 enables the ratio C/F to be calculated.

The rate of release of ^3H into the cell water T is given by the equation

$$T = \text{net flux of } ^3H \text{ to triose phosphate}$$

which, since the specific radioactivity of triose phosphates is zero,

$$= \text{net flux of } ^3H \text{ to fructose diphosphate}$$

$$= F \cdot s_p - C \cdot s_f$$

so that

$$T/s_g = F \cdot s_1 - C \cdot s_2$$

and

$$F = (T/s_g + C \cdot s_2)/s_1 \qquad (2)$$

Eqns. (1) and (2) can be combined to eliminate F, so that the rate of cycling C is given by the equation

$$C = \frac{T(1 - s_1)}{s_g(s_1 - s_2)}$$

Consequently, the determination of T, s_1 and s_2 together with the specific radioactivity of the administered glucose (relative to ^3H) enables the rate of cycling to be calculated. Eqn. (1) may then be used to calculate the rate of the phosphofructokinase reaction F and hence the glycolytic flux ($F - C$).

The experimental determination of s_1 and s_2 is simplified by the use of ^{14}C as an internal standard for s_g (i.e. by administering [U-^{14}C,5-^3H]glucose). Since there is no loss or gain of carbon atoms in the system, the specific radioactivity (relative to ^{14}C) of each metabolite is the same as that of glucose. Consequently, the ^{14}C radioactivity of each metabolite is proportional to s_g, and s_1 and s_2 can be calculated directly from the $^3H/^{14}C$ radioactivity ratios of glucose, hexose phosphates and fructose diphosphate. For example,

$$s_1 = s_p/s_g$$

$$= \frac{^3H \text{ radioactivity of a given concentration of hexose phosphate}}{^3H \text{ radioactivity of an equal concentration of glucose}}$$

Since the ^{14}C specific radioactivities of hexose phosphates and glucose are equal,

$$s_1 = \frac{^3H/^{14}C \text{ ratio of hexose phosphate}}{^3H/^{14}C \text{ ratio of glucose}}$$

Similarly,

$$s_2 = \frac{{}^3H/{}^{14}C \text{ ratio of fructose diphosphate}}{{}^3H/{}^{14}C \text{ ratio of glucose}}$$

It should be noted that the application of this method depends on the near-equilibrium nature of aldolase. If this reaction was non-equilibrium there would be no dilution of the specific radioactivity of fructose diphosphate by triose phosphates, the values of s_1 and s_2 would both be unity and the equation for C would be indeterminate. At the other extreme, total equilibrium at aldolase would result in a zero value for s_f and hence s_2; this latter condition has been assumed by Bloxham *et al.* (1973), on the basis of evidence from studies with a reconstituted system *in vitro*.

Finally, this method for measuring the rate of substrate cycling between fructose 6-phosphate and fructose diphosphate depends on the absence of a significant activity of the pentose phosphate cycle. Consequently, it must be used with appropriate caution in tissues other than muscle (see Katz & Rognstad, 1975).

References

Bloxham, D. P., Clark, M. G., Holland, P. C. & Lardy, H. A. (1973) *Biochem. J.* **134**, 584–587
Clark, D. G., Rognstad, R. & Katz, J. (1973) *Biochem. Biophys. Res. Commun.* **54**, 1141–1148
Katz, J. & Rognstad, R. (1975) *Curr. Top. Cell. Regul.* in the press

Biochem. Soc. Symp. (1976) **41**, 111–131
Printed in Great Britain

Biochemical Adaptations for Flight in the Insect

By BERTRAM SACKTOR

Gerontology Research Center, National Institute on Aging, National Institutes of Health, Baltimore City Hospitals, Baltimore, MD 21224, U.S.A.

Synopsis

1. Flight by insects is characterized by the most intense respiration known in biology and also the most controlled. Thus insect flight muscle may be the tissue of choice for the study of biochemical adaptation in the control of catabolism and biological oxidations, and many of the results obtained with insects have a significance and a relevance that transcend the boundaries between classes. In insects, such as the blowfly, flight is distinguished additionally by high wingbeat frequencies and an asynchronous type of excitation–contraction coupling. In spite of this intense muscular work, metabolic processes are not limited by the availability of oxygen. Also of importance is the morphological organization of the flight muscle and mitochondria, which have evolved ultrastructurally and biochemically into an effective catabolic machine. 2. In the fly, carbohydrate, principally glycogen, is the sole metabolic fuel; fats are not used in flight and enzymes concerned with fatty acid utilization are virtually lacking. Glycogenolysis does not lead to lactic acid; instead, the end products of glycolysis are pyruvate and α-glycerophosphate. The α-glycerophosphate cycle provides a mechanism not only for the reoxidation of glycolytically produced NADH but also for the stoicheiometric formation from each molecule of hexose equivalent of two molecules of pyruvate, which are then available for oxidation via the tricarboxylate cycle. The absence of dicarboxylate and tricarboxylate carriers from the mitochondria ensures that tricarboxylate-cycle intermediates do not exit from the mitochondrion but that pyruvate is oxidized to completion. On initiation of flight, mitochondrial oxidation of pyruvate is impeded by the lack of tricarboxylate-cycle intermediates for the generation of oxaloacetate. This is circumvented by the oxidation of proline. 3. The controls on metabolism in flight muscle, i.e. (1) glycogenolysis at phosphorylase and phosphorylase kinase, (2) glycolysis at phosphofructo-kinase, (3) α-glycerophosphate dehydrogenase, (4) proline dehydrogenase and (5) tricarboxylate cycle at isocitrate dehydrogenase, are effected by the phosphate potential and/or Ca^{2+}. It is suggested that the metabolic changes, such as those seen in the rest-to-flight transition, are achieved by the concerted actions of these effectors at the different loci.

The intensity and precise control of metabolic processes in flight muscle of insects have engaged the attention of biochemists and physiologists for many years. To illustrate, metabolic rates as high as 2400cal/h per g of muscle during prolonged periods of continuous flight of the bee have been reported (Weis-Fogh, 1952). This value is 30- to 50-fold those for leg and heart muscle of man at maximum activity. Additionally, on initiation of flight some insects, such as the blowfly, respire at a rate of $3000\,\mu l$ of oxygen/min per g, elevating their basal rates approximately 100-fold (Davis & Fraenkel, 1940). This is the most intense respiration known in biology and also the most controlled. From these selected examples, and others reported elsewhere (Sacktor, 1965, 1970), it is apparent that insect flight muscle is the tissue of choice for the study of biochemical adaptation in the control of catabolism and biological oxidations, and many of the results obtained with insects have a significance and a relevance which transcend boundaries between classes.

The metabolic rates cited above are typical of those found in flies, such as the blowfly. In these insects flight is characterized by an asynchronous type of excitation–contraction coupling and by high wingbeat frequencies. Such large increases in respiration on initiation of flight are not restricted to the *Diptera* and *Hymenoptera*, however. Essentially identical increases in oxygen uptake between individuals at rest and during flight have been observed in locusts and moths, which have a synchronous type of excitation–contraction coupling and, in general, have relatively low rates of wingbeat (cf. Sacktor, 1970). In spite of this intense muscular work, insects can maintain flight for hours while accruing little, if any, oxygen debt. This indicates that the metabolic processes are not limited by the availability of oxygen. In insects, air is conveyed directly to the flight muscle through an elaborate conduit of tracheae, which invade the fibres and are in close juxtaposition with each mitochondrion. Furthermore, Weis-Fogh (1964, 1967) has calculated that in many insects (e.g. flies) diffusion of respiratory gases is sufficient to account for the entire transport between the spiracle and the end of the tracheole, even at the highest rate of metabolism.

Also of paramount importance to the understanding of adaptations for flight is the morphological and ultrastructural organization of the flight muscle and mitochondrion, shown in Plate 1. Included among the distinguishing features are the large fibrils, a markedly decreased sarcoplasmic reticulum, ample stores of glycogen and relatively enormous mitochondria that comprise 40% of the muscle mass and half of the total protein. The mitochondria are tightly packed with numerous cristae that nearly fill the entire lumen. This intramitochondrial arrangement differs significantly from that found in other kinds of mitochondria, i.e. those from mammalian liver. In rat liver, the cristae are sparse and irregular, and the inner compartment (matrix) is large. The differences between the ultrastructure of mitochondria from blowfly flight muscle and those from rat liver reflect differences in the known function of the two organelles. Flight-muscle mitochondria have a requirement for a large number of respiratory assemblies corresponding to the intense respiratory capability of the mitochondria that is expressed during flight. In contrast, mitochondria from liver cells need fewer respiratory assemblies corresponding to their lower rates of oxidation. On the other hand, neither the biosynthetic activities nor the urea-cycle enzymes of rat liver mitochondria are present in blowfly flight-muscle mitochondria; thus there is less need for matrix space. These comparisons

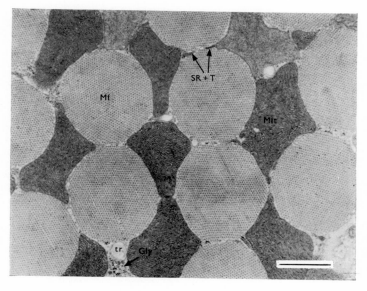

EXPLANATION OF PLATE I

*An electron micrograph of blowfly Phormia regina flight muscle in cross-section showing arrange-
ment of mitochondria (Mit) and myofibrils (Mf)*

Other membranous structures evident are elements of the sarcoplasmic reticulum (SR) and the
T-system (T). The T-system is comprised, in part, of elements of the plasma membrane, which
surround the invaginating tracheoles (tr.). Glycogen rosettes (Gly) fill the sarcoplasmic spaces.
The bar represents 1 μm. This is taken from Reed & Sacktor (1971).

B. SACKTOR

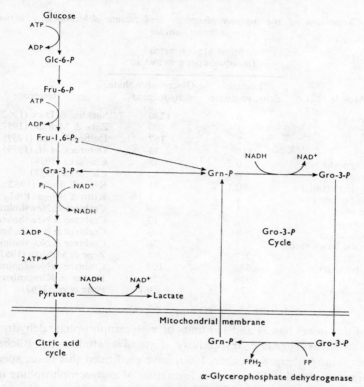

Fig. 1. *Schematic diagram of glycolysis and the α-glycerophosphate cycle in flight muscle*

This is taken from Sacktor (1970).

suggest that flight muscle has evolved ultrastructurally and biochemically into an effective catabolic machine.

In the fly, carbohydrate, principally glycogen and trehalose, is the sole metabolic fuel (Sacktor, 1975). Fats are not used in flight and enzymes concerned with fatty acid utilization are virtually lacking. Glycolysis in flight muscle differs significantly from that found in mammalian muscle. As illustrated diagrammatically in Fig. 1, the NADH formed by glyceraldehyde phosphate dehydrogenase is reoxidized concomitantly with a conversion of dihydroxyacetone phosphate into α-glycerophosphate, catalysed by the cytosolic α-glycerophosphate dehydrogenase. Pyruvate is not converted into lactate; indeed in the fly lactate dehydrogenase is essentially absent from flight muscle (Sacktor, 1955).

The glycolytic pathway as described in Fig. 1, in which α-glycerophosphate and pyruvate are the end products, is characteristic of the flight muscle of insects, especially those with high-frequency wingbeat. However, marked differences are found in this respect between tissues in a given species and between the same tissue in different orders of insects. For example, as shown in Table 1, flight muscle of the locust has 167 units of α-glycerophosphate dehydrogenase and only 2 units of lactate dehydrogenase. In contrast, the leg

Table 1. *Comparison of the α-glycerophosphate and lactate dehydrogenase activities in different muscles*

| | Substrate converted (μmol/min per g wet wt.) | | |
| | Lactate dehydrogenase | α-Glycerophosphate dehydrogenase | |
Muscle	Lactate dehydrogenase	α-Glycerophosphate dehydrogenase	Reference
Flight (blowfly)	0	1230	Sacktor & Dick (1962)
Flight (bee)	3	700	Zebe & McShan (1957)
Flight (locust)	2	167	Delbruck et al. (1959)
Leg (locust)	117	33	Delbruck et al. (1959)
Flight (cockroach)	0.2	48	Chefurka (1958)
Leg (cockroach)	0.1	32	Chefurka (1958)
Flight (praying mantis)	<0.1	11	Kitto & Briggs (1962a,b)
Leg (praying mantis)	1	1	Kitto & Briggs (1962a,b)
Flight (waterbug)	1	51	Crabtree & Newsholme (1972)
Leg (waterbug)	59	13	Crabtree & Newsholme (1972)
Flight (cockchafer)	4	103	Crabtree & Newsholme (1972)
Flight (poplar hawk moth)	3	36	Crabtree & Newsholme (1972)
Tail (crayfish)	217	5	Zebe & McShan (1957)
Pectoral (pheasant)	542	103	Crabtree & Newsholme (1972)
Skeletal (rat)	330	50	Bücher & Klingenberg (1958)
Smooth (bovine)	25	0.1	Pette et al. (1962)

muscle of this insect has 33 and 117 units of α-glycerophosphate dehydrogenase and lactate dehydrogenase respectively. Extensive studies by Bücher and colleagues (Klingenberg & Bücher, 1960) have confirmed this tissue specificity. They point out that the characteristic formation of α-glycerophosphate in flight muscle is in accord with its aerobic nature. In contrast, lactate is produced during the reoxidation of NADH in those muscles that may become temporarily anoxic. Of additional interest in comparing the activities of the two dehydrogenases is the observation of Brosemer (1967) on the flightless grasshopper *Romalea microptera*. The wing muscle of this insect, which during evolutionary development has lost the ability to fly, no longer has the biochemical characteristics of flight muscle; i.e. high α-glycerophosphate dehydrogenase and low lactate dehydrogenase. Instead, the activities of the dehydrogenases in the wing muscles are similar to those found in the insect's leg muscles.

In insect flight muscle, the α-glycerophosphate formed by the cytosolic α-glycerophosphate dehydrogenase is readily accessible to the mitochondrial α-glycerophosphate dehydrogenase and is oxidized, in turn, by the flavoprotein, thereby regenerating dihydroxyacetone phosphate. This triose phosphate is then available for further oxidation of extramitochondrial NADH. Accordingly, the two reactions, i.e. the reduction of dihydroxyacetone phosphate and the oxidation of α-glycerophosphate, constitute the α-glycerophosphate cycle (Estabrook & Sacktor, 1958; Bücher & Klingenberg, 1958). As illustrated in Fig. 1, the cycle is a shuttle system in which cytosolic NAD^+-linked substrates enter and leave the mitochondria in reduced and oxidized states respectively. In this way, reducing equivalents from the extramitochondrial pool of NADH pass the cytosol–mitochondrial permeability barrier and are oxidized. At the same time, reduced flavoprotein in the mitochondria donates its reducing equivalents to the electron-transport chain. Further, the cyclic process is self-generating, only a catalytic quantity of dihydroxyacetone phosphate being needed to oxidize the continuously formed NADH (Sacktor & Dick, 1962). This suggests that most of the dihydroxyacetone phosphate produced in the aldolase reaction can be isomerized to glyceraldehyde phosphate. Also, all the carbon derived from the carbohydrate metabolized during prolonged flight can be converted into pyruvate, and hence is available for oxidation in the mitochondria via the tricarboxylate cycle. This is evident from findings that there is no accumulation of partially oxidized intermediates (α-glycerophosphate, pyruvate, lactate and alanine) during flights of 1h or more by the blowfly (Sacktor & Wormser-Shavit, 1966) and the locust (Kirsten et al., 1963). At shorter time-intervals accumulation of some of these metabolites does occur and may be of significance (Sacktor & Wormser-Shavit, 1966; Childress et al., 1967). However, these studies in vivo demonstrate that after an initial lapse, the mitochondria are capable of oxidizing pyruvate and α-glycerophosphate at the rates in which they are formed.

The large deposits of glycogen in flight muscle as well as the depletion of these reserves during flight indicate that in many insects, especially Diptera, glycogen provides a major vehicle for storage of flight energy which is rapidly mobilized to meet the metabolic requirements of the muscle. For example, in the blowfly Phormia regina fed ad lib, glycogen comprises 10–15mg/g wet wt. of the thorax (Childress et al., 1970). During flight, approximately 2.5 μmol (as equivalents of glucose)/g wet wt. are utilized each minute until the polysaccharide is depleted (Sacktor & Wormser-Shavit, 1966). This rapid utilization of glycogen in muscle, commencing after flight is induced, indicates control of glycogenolysis. Glycogen phosphorylase, in both a and b forms, from flight muscle of the blowfly P. regina has been purified to a high degree of homogeneity and the kinetics of the enzymes have been characterized (Childress & Sacktor, 1970). It has been estimated that phosphorylase comprises approximately 1.5% of the total muscle protein and has a potential activity of 9.6 μmol/min per g wet wt. of thorax, a value more than adequate to account for the rate of glycogenolysis during flight.

The kinetic properties of the purified phosphorylases a and b from blowfly flight muscle, including the interactions between the co-substrates, glycogen and P_i, activator, AMP and inhibitor, ATP with the enzymes have been described (Childress & Sacktor, 1970). Double-reciprocal plots of the activity of either phosphorylase a or b and concentration of substrate, at different concentrations of co-substrate, indicate a mechanism in which increasing concentrations of either substrate enhance the binding of the other. Although not required for activity, AMP stimulates phosphorylase a 2–3-fold at saturating concentrations of substrates and 10-fold or higher at low substrate concentrations. The apparent K_m for AMP is 0.6 μM at saturating concentrations of phosphate and glycogen. Lowering the concentration of either substrate decreases the affinity of the enzyme for the activator. Moreover, AMP increases the affinity of phosphorylase a for both substrates. For example, the apparent K_m for phosphate is lowered from 100 to 9mM in the presence of low and high concentrations of AMP respectively. Unlike the a form of the enzyme, phosphorylase b has an absolute requirement for AMP. Furthermore, concentrations of AMP 100-fold greater than those which stimulate phosphorylase a are needed to stimulate phosphorylase b.

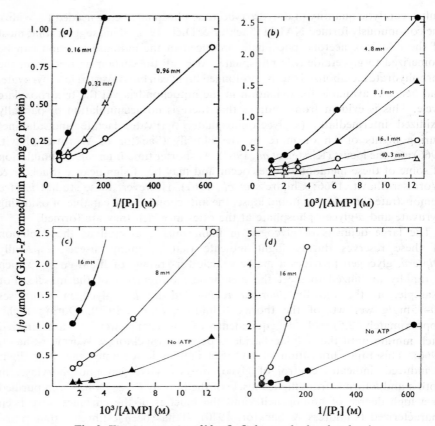

Fig. 2. *Kinetic properties of blowfly flight-muscle phosphorylase b*

(*a*) Double-reciprocal plots of initial velocity of phosphorylase *b* as a function of P_i concentration at several concentrations of AMP (marked on Figure). The concentration of glycogen is 2mM. (*b*) Double-reciprocal plots of initial velocity of the enzyme as a function of AMP concentration at several concentrations of P_i (marked on Figure). Glycogen concentration is 2mM. (*c*) Double-reciprocal plots showing the effect of ATP on the initial activity of phosphorylase *b* as a function of AMP concentrations. Substrate concentrations are 16mM-P_i and 2mM-glycogen. (*d*) Double-reciprocal plots showing the effect of ATP on the initial activity of phosphorylase *b* as a function of P_i concentration. The concentrations of AMP and glycogen are 0.32mM and 2mM respectively. (From Childress & Sacktor, 1970.)

Increasing amounts of AMP lower the apparent K_m values for both phosphate and glycogen. As illustrated in Fig. 2, the apparent K_m values for phosphate range from 100mM at the lowest to 5mM at the highest concentration of AMP. Increasing concentrations of glycogen or phosphate lower the affinity of phosphorylase *b* for AMP. It has been estimated from the data in Fig. 2 that the apparent K_m values for AMP decrease from 2 to 0.2mM as the phosphate concentration increases from 4.8 to 40mM. Additionally, Childress & Sacktor (1970) have found that ATP is a potent inhibitor of phosphorylase *b* but not of phosphorylase *a*. The K_i value is approximately 2mM (Fig. 2). At high AMP concentrations the inhibition by ATP is less, suggesting competitive inhibition with respect to AMP. This view is strengthened by the observations that ATP

Table 2. *Comparison of metabolite concentrations and apparent K_m values for flight-muscle phosphorylases under conditions in vivo*

The results are taken from Childress & Sacktor (1970). The apparent K_m values were determined from the kinetic data with the concentrations of co-substrate and activator as found *in vivo* at rest. Potential activities are expressed as the percentage of potential phosphorylase activity with saturating conditions of substrate and AMP, without ATP.

Metabolite	Concentration at rest in vivo (μmol/g wet wt. of thorax)	Concentration during flight in vivo (μmol/g wet wt. of thorax)	Apparent K_m value (mM)		Potential activity (%) at simulated conditions of			
					Rest		Flight	
			b	a	b	a	b	a
Glycogen	6–9	6→0.75*	0.6	1.7	—	—	—	—
P_i	7.0	7.5	100	8	—	—	—	—
AMP	0.1	0.3	1.0	0.001	0	50	0	50
ATP	7.0	6.5	$2(K_i)$	—	—	—	—	—

* The glycogen concentration in the muscle *in vivo* decreased steadily during flight to a value of 0.75 mM after 10 min of flight. It remained steady at this exhausted level throughout the remainder of the flight. A glycogen concentration of 1% corresponds to 5.9 mM end groups.

increases the apparent K_m values for both glycogen and phosphate, as if AMP has been displaced from its binding site on the enzyme.

These detailed kinetic data for blowfly flight-muscle phosphorylases, coupled with a knowledge of the concentrations of the substrates, activator and inhibitor in the muscle during rest and flight (Sacktor & Wormser-Shavit, 1966; Sacktor & Hurlbut, 1966), permit estimates of the phosphorylase activities under simulated conditions *in vivo*. As shown in Table 2, the concentration of glycogen in the muscle is sufficient to saturate the enzymes, except after 10 min or more of flight when the muscle glycogen reserve is near depletion. The concentration of phosphate in the muscle at rest is 7.0 mM, increasing to 7.5 mM during flight. On the other hand, the concentration of AMP in the muscle is increased sharply from 0.1 to 0.3 mM in the rest-to-flight transition. However, examination of the apparent K_m values for phosphate and AMP reveals that each ligand has a marked effect on the affinity of the enzyme for the other in these concentration ranges. For phosphorylase *b*, the apparent K_m for phosphate at 0.1 mM-AMP is about 100 mM, very much above that found in the muscle *in vivo*. The apparent K_m of phosphorylase *b* for AMP at 8 mM-phosphate is about 1.0 mM, about 10-fold that in the muscle at rest. These observations indicate that the activity of the *b* form would be limited to only a trace of its potential activity at substrate saturation. In addition, the strong inhibitory effect of ATP on phosphorylase *b*, at the low concentration of AMP in the tissue, would decrease the activity of phosphorylase *b* to almost zero. In contrast, for phosphorylase *a*, the concentration of AMP in the muscle is 100-fold the apparent K_m of the nucleotide. Furthermore, ATP does not inhibit the *a* form of the enzyme. The apparent K_m of phosphorylase *a* for phosphate, with conditions simulating those *in vivo*, is 8 mM, a value approximating to the concentration in the muscle. Thus about 50% of the potential activity of phosphorylase *a* would be found and glycogenolysis by phosphorylase *a* would be moderately responsive to changes in the concentration of phosphate.

Table 3. *Relative amounts of phosphorylases a and b in vivo at rest and during flight*

The values are taken from Childress & Sacktor (1970). They are means±s.e.m. of the numbers of measurements given in parentheses. Mounted-rested means

State	Phosphorylase in the *a* form (% of total)
Resting	17.8 ± 3.6 (17)
Flown 5 s	63.9 ± 12.3 (16)
Flown 15 s	69.1 ± 13.4 (16)
Flown 30 s	72.2 ± 12.3 (10)
Flown 60 s	71.5 ± 14.2 (10)
Flown 10 min	72.1 ± 13.1 (10)

On the basis of a specific activity for flight-muscle phosphorylase of 9.6 units per g wet wt. of thorax (Childress *et al.*, 1970) and the fact that the potential activity of phosphorylase is the same whether it exists in the *a* or the *b* form, it can be estimated from the kinetic data that at least 50% of the total phosphorylase must be in the *a* form in order to account for the rate of glycogen breakdown that occurs during flight, 2.4 μmol of glucosyl residues/ min per g wet wt. of thorax (Sacktor & Wormser-Shavit, 1966). It is suggested, therefore, that to satisfy this rate of glycogenolysis, phosphorylase *b* must be converted into phosphorylase *a* on initiation of flight. This view is supported by direct measurements of two forms of phosphorylase in the muscle, during rest and during flight (Childress & Sacktor, 1970). As shown in Table 3, in 'resting' blowflies 18% of the phosphorylase is in the *a* form. Initiation of flight induces an immediate increase in the relative amount of the enzyme in the *a* form, reaching a maximum of about 70% at 15 s of flight. The total amount of phosphorylase present (*a* plus *b*) does not change during flight. This level of phosphorylase *a* is adequate to account for the observed rate of glycogenolysis during flight. From the kinetic data obtained under conditions *in vitro*, a 70% level of phosphorylase *a* can catalyse a rate of glycogenolysis of 4.3 μmol of glucosyl residues/min per g wet wt. of thorax as compared with the rate of 2.4 actually measured. If, in 'resting' blowflies, the value of 18% of the total phosphorylase in the *a* form is too high because of technical difficulties in extracting the phosphorylases in their state *in situ*, and if these factors increase the amount of phosphorylase *a* in the flown fly to the same extent, the agreement between the predicted rate of glycogenolysis from studies *in vitro* and that in the flying insect would be even more exact.

The conversion of phosphorylase *b* into phosphorylase *a* is catalysed by the enzyme phosphorylase *b* kinase. The mechanisms by which phosphorylase *b* kinase becomes activated are, therefore, of significance to the regulation of flight-muscle metabolism. Hansford & Sacktor (1970*a*) have found that phosphate and Ca^{2+}, at physiological concentrations, activate phosphorylase *b* kinase. Stimulation of the enzyme is evident at concentrations of Ca^{2+} as low as 10 nM, with maximal enhancement at about 1 μM. Mammalian skeletal-muscle phosphorylase *b* kinase is also activated by Ca^{2+}. However, comparative studies show that the responses of the enzymes from the two tissues to

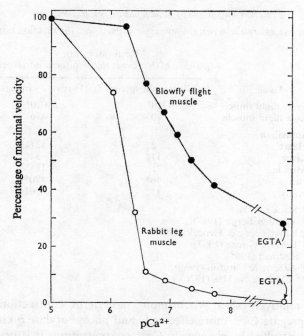

Fig. 3. *A comparison of the effect of concentration of Ca²⁺ on the activities of phosphorylase b kinases from blowfly flight muscle and rabbit leg muscle*

The points EGTA [ethanedioxybis(ethylamine)tetra-acetic acid] on the plots correspond to no added CaCl₂. From Sacktor *et al.* (1974). pCa^{2+} is $-\log[Ca^{+2}]$.

Ca²⁺ are different (Sacktor *et al.*, 1974). As illustrated in Fig. 3, rabbit skeletal-muscle kinase has virtually no activity in the absence of Ca²⁺ and exhibits a large increase in activity over a narrow range of Ca²⁺ concentrations. The phosphorylase *b* kinase from blowfly flight muscle has appreciable activity in the absence of exogenous Ca²⁺ and has a smaller increase, only threefold, over a wide range of Ca²⁺ concentration. The concentrations of Ca²⁺ required for half-activation are 0.1 and 1 μM for the blowfly and rabbit enzymes respectively.

The activation of phosphorylase *b* kinase by Ca²⁺ provides support for the concept that in muscle the conversion of phosphorylase *b* into *a* represents an essential component in the mechanism coupling contraction to glycogenolysis. The low Ca²⁺ sensitivity of the kinase in blowfly flight muscle may be of additional physiological importance. Flight muscle of the blowfly shows contraction–relaxation cycles at a constant Ca²⁺ concentration (Jewell & Ruegg, 1966), and Ca²⁺ concentration can be varied considerably without loss of oscillatory work (Pringle & Tregear, 1969). Moreover, blowfly muscle has very little sarcoplasmic reticulum (Plate 1) and it appears that rapid segregation of Ca²⁺ within membrane-bounded vesicles is not required for oscillatory behaviour of the asynchronous flight muscle. Therefore a high Ca²⁺-sensitivity for phosphorylase *b* kinase would be disadvantageous. On the other hand, muscles having a well-developed sarcoplasmic reticulum and

5

Table 4. *Comparison of mitochondrial respiratory rates*

With mammalian mitochondria, when the substrate is pyruvate, the reaction mixture contains malate.

	Respiratory rate (ng-atoms of O/min per mg of mitochondrial protein)	
Tissue	α-Glycerophosphate	Pyruvate+(malate)
Blowfly flight muscle	4320*	4020*
Locust flight muscle	280†	476†
Mammalian		
Heart	2‡	121§
Liver	11‡	34‡
Muscle	71‖	185¶
Brain	46**	170**
Kidney	11‡	50‡

* Slack (1975).
† Van den Bergh (1967).
‡ Klingenberg & Slenczka (1959).
§ Tyler & Gonze (1967).
‖ Hedman (1965).
¶ Azzone & Carafoli (1960).
** Clark & Nicklas (1970).

possessing a relatively slow synchronous pattern of contraction–relaxation, are able to segregate Ca^{2+} more effectively and phosphorylase *b* kinase activity can be fully controlled by changes in Ca^{2+} concentration. It is probably more than coincidental that the distinction between asynchronous and synchronous types of muscle in the Ca^{2+} sensitivities of their phosphorylase *b* kinases has a precise counterpart in the distinction between asynchronous and synchronous muscles in the Ca^{2+} sensitivities of their actomyosin ATPases* (Maruyama *et al.*, 1968). It is also noteworthy that no evidence has been found for a cyclic AMP-dependent protein kinase in flight muscle to convert phosphorylase *b* kinase into an active form. Apparently only Ca^{2+} and phosphate are necessary.

As discussed above, glycolysis in flight muscle of the blowfly, and in other species as well to varying degrees, gives rise to α-glycerophosphate and pyruvate. Since the end products of glycolysis do not accumulate during prolonged flights, the exceptional capacity of blowfly mitochondria to oxidize α-glycerophosphate and pyruvate is indicated. That this is indeed the case is shown in Table 4, which compares the rates of oxidation of these two substrates in blowfly and locust flight muscle with those in mammalian tissues. Values as high as 4000 ng-atoms of oxygen/min per mg of protein have recently been reported for the oxidation of α-glycerophosphate and pyruvate by mitochondria of the blowfly *Sarcophaga nodosa* (Slack, 1975). These values are severalfold those of locust flight-muscle mitochondria and several orders of magnitude greater than those found for mitochondria from mammalian tissues. Oxidative rates of 1000–2500 ng-atoms of oxygen/min per mg of protein for other species of blowflies have previously been reported (Hansford & Sacktor, 1971).

When different substrates are added to a suspension of mitochondria isolated from flight muscle of flies, rates of oxidation as shown in Table 5 are obtained. It is evident that intact mitochondria oxidize, at appreciable rates, only

* Abbreviation: ATPase, adenosine triphosphatase.

Table 5. *Respiratory activities of mitochondria from blowfly flight muscle*

Data compiled from Sacktor & Childress (1967), Childress *et al.* (1967), Childress & Sacktor (1966), Chance & Sacktor (1958) and Bulos *et al.* (1972).

Substrate	Rate (μg-atoms of O/min per mg of mitochondrial protein)
α-Glycerophosphate	1.35
Pyruvate	0.98
Acetylcarnitine	0.49
Proline	0.19
Citrate	0.01
α-Oxoglutarate	0.07
Succinate	0.09
Fumarate	0.05
Malate	0.04
Glutamate	0.04
Aspartate	0.01

exogenous α-glycerophosphate, pyruvate, acetylcarnitine and, to a lesser extent, proline. In contrast, intermediates of the tricarboxylate cycle, i.e. citrate, isocitrate, α-oxoglutarate, succinate, fumarate and malate, the amino acids glutamate and aspartate, and NADH added to isolated mitochondria, are not effective respiratory substrates. An explanation for the low rates of oxidation of intermediates of the tricarboxylate cycle and acidic amino acids stems from the discovery of Van den Bergh & Slater (1962) that flight-muscle mitochondria are not readily permeable to these compounds. Subjecting mitochondria to sonic disintegration or freezing–thawing, procedures that disrupt the integrity of the mitochondrial membranes, increases the respiratory rates with these substrates manyfold (Van den Bergh & Slater, 1962; Van den Bergh, 1967; Sacktor & Childress, 1967). On the other hand, the oxidations of α-glycerophosphate and pyruvate are not stimulated, and that of proline may be markedly decreased by these and other disruptive procedures (Sacktor & Childress, 1967; Norden & Ventouris, 1972). Other techniques, e.g. ammonium salt-swelling tests (Chappell & Crofts, 1966), also indicate that intermediates of the tricarboxylate cycle do not penetrate the inner membrane of flight-muscle mitochondria. Thus these experiments suggest the absence from flight-muscle mitochondria of the specific dicarboxylate and tricarboxylate anion permeases for the exchange-diffusion of tricarboxylate-cycle substrates, as described in mammalian mitochondria. Not only are these intermediates unable to enter the mitochondrion but, perhaps, more importantly they cannot leave. This compels their complete oxidation to CO_2 and water.

As shown in Table 5 and as discovered earlier (Sacktor, 1955), mitochondria isolated from flight muscle of flies oxidize proline at a moderately high rate. The oxidation of proline by flight-muscle preparations has now been reported for all the major orders of insects (Crabtree & Newsholme, 1970). The physiological significance of the oxidation of proline in flight muscle of the blowfly, particularly at the initiation of flight, is apparently the provision of tricarboxylic acid-cycle intermediates needed for the maximal rate of oxidation of pyruvate (Sacktor & Wormser-Shavit, 1966; Childress & Sacktor, 1966; Sacktor & Childress, 1967). This suggestion is supported by experiments

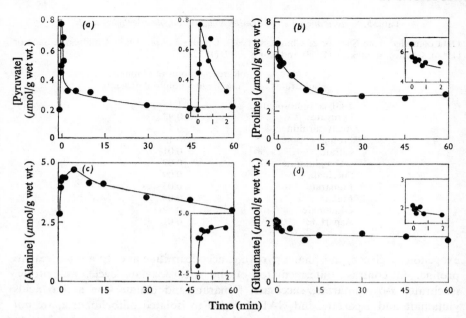

Fig. 4. *Sequential changes in the concentration of pyruvate* (a), *proline* (b), *alanine* (c) *and glutamate* (d) *in flight muscle of the blowfly during* 1 h *of flight*

This is taken from Sacktor & Wormser-Shavit (1966). The insets show more clearly the changes in metabolite concentration during the first 2 min of flight.

showing that isolated mitochondria rapidly lose their ability to oxidize pyruvate and that this loss is reversed by proline, but not by tricarboxylate-cycle intermediates nor glutamate. This finding is consistent with the ability of blow-fly mitochondria to oxidize exogenous proline, but not the other metabolites, when added exogenously (Table 5). Proline is converted into glutamate by the actions of proline and Δ'-pyrroline-5-carboxylate dehydrogenases. Transamination of glutamate with pyruvate gives rise to alanine and α-oxoglutarate. The intramitochondrial α-oxoglutarate is further metabolized via the tricarboxylate cycle to form oxaloacetate. This oxaloacetate can then condense with acetyl-CoA, derived from pyruvate, forming citrate and thus effecting the complete oxidation of pyruvate by means of the tricarboxylic acid cycle. This concept receives additional substantiation from studies of metabolite concentrations during rest and flight of the blowfly (Sacktor & Wormset-Shavit, 1966), as shown in Fig. 4. During the first few seconds of flight the concentration of proline in flight muscle decreases abruptly. At the same time there is a stoicheiometric accumulation of alanine, suggesting that it is only when proline oxidation has yielded sufficient oxaloacetate does pyruvate begin to be oxidized as rapidly as it is being produced by glycolysis. In the interim, pyruvate transaminates with glutamate to form alanine (Sacktor & Childress, 1967), or forms acetyl-CoA, which gives rise to acetylcarnitine (Childress *et al.*, 1967) or is deacylated to yield acetate (Tulp & Van Dam, 1970).

A definitive explanation for the unique role of proline in providing precursors of oxaloacetate has yet to be formulated. Two possibilities seem attractive at this time. One is that proline, having no net charge, can pass through the mitochondrial membrane permeability barrier, whereas glutamate, aspartate and citric acid-cycle intermediates, carrying net charges, are unable to penetrate the barrier in the absence of specific carriers. The second possibility is that proline dehydrogenase is located on the outer surface of the inner mitochondrial membrane. Thus, like α-glycerophosphate, there would be no barrier to proline. This hypothesis receives some support from the observation that ADP activates proline dehydrogenase in the presence of atractyloside (Hansford & Sacktor, 1970b), suggesting that the action of ADP on the dehydrogenase may be external to the ADP translocase, an inner-membrane enzyme.

Although the suggested physiological importance of the oxidation of proline in flight muscle of the blowfly is the provision of oxaloacetate on the initiation of flight, the oxidation of proline may play a larger role in flight-muscle metabolism of the Tsetse fly. In an elegant series of papers Bursell (1963, 1965, 1966, 1967) has shown that the concentration of proline decreases markedly during flight and that the content of alanine rises concomitantly and nearly stoicheiometrically. He proposes a metabolic cycle, which brings about the conversion of proline into alanine, CO_2, NADH and reduced flavin. The scheme envisages the use of only a segment of the tricarboxylate cycle, from α-oxoglutarate to oxaloacetate. A calculation reveals that the number of molecules of ATP that are produced by the conversion of 1 mol of proline into alanine by this scheme is 14, a value comparing favourably with the 15 mol of ATP produced by the complete oxidation of 1 mol of pyruvate via the citric acid cycle.

Remarkably high activities of proline dehydrogenase have also been found in the flight muscle of beetles, suggesting that the oxidation of proline has a greater and more widespread relevance than previously supposed. In the Colorado potato beetle, proline is oxidized at a rate severalfold those of pyruvate and α-glycerophosphate (de Kort et al., 1973). The highest rate of oxygen consumption is also obtained with proline in the Japanese beetle (Hansford & Johnson, 1975). In the cockchafer, the activity of proline dehydrogenase is similar in magnitude to the activities of two important enzymes of the tricarboxylate cycle, i.e. NAD^+-linked isocitrate dehydrogenase and succinate dehydrogenase (Crabtree & Newsholme, 1970). In the Japanese beetle, however, the oxidation of proline leads to the production of more NH_3 than alanine, indicating a functioning glutamate dehydrogenase. Studies of mitochondrial extracts confirm the presence of a very active glutamate dehydrogenase as well as a uniquely active 'malic' enzyme. Hansford & Johnson (1975) suggest that the input of tricarboxylate-cycle intermediate from proline oxidation is balanced by the formation of pyruvate from malate, followed by the complete oxidation of the pyruvate.

It has been pointed out that the large increase in the rate of oxygen uptake on initiation of flight, over 100 times in some cases, indicates that there is an exceptionally high degree of respiratory control in flight-muscle mitochondria, in situ. In the discussion that follows, studies suggesting possible mechanisms contributing to this regulation will be examined.

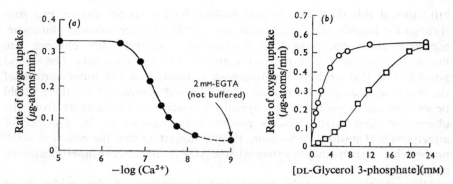

Fig. 5. *Control of α-glycerophosphate oxidation of Ca²⁺ with mitochondria isolated from flight muscle of the blowfly Calliphora vomitora*

(*a*) Effect of the concentration of free Ca^{2+}, established with $EGTA$–Ca^{2+} buffers, on the rate of oxidation. (*b*) Effect of the concentration of α-glycerophosphate on the rate of oxidation in the presence (○) and absence (□) of Ca^{2+}. From Hansford & Chappell (1967).

As described previously α-glycerophosphate is one of the end products of glycolysis in flight muscle. Sacktor & Wormser-Shavit (1966) have found that concomitant with the increased glycolytic flux associated with flight the mito-chondrial oxidation of α-glycerophosphate is activated and have suggested that this activation represents one of the biochemical control points in the metabolism of the muscle.

In early studies, Chance & Sacktor (1958) have hypothesized that the mode of control of α-glycerophosphate oxidation in flight muscle may be novel, in that the dehydrogenase rather than the respiratory chain is limiting. Subsequently, Estabrook & Sacktor (1958) have found that bivalent-cation chelators inhibit α-glycerophosphate dehydrogenase and that this inhibition is reversed by addition of Ca^{2+} or Mg^{2+} and by excess of substrate. From these and additional findings it has been suggested that activation of α-glycerophosphate involves the reversal, coincident with the initiation of flight, of an inhibited resting state, effected by the release of bivalent cation from a sequestering system in the muscle, i.e. the sarcoplasmic reticulum. This proposal receives strong support from the investigations of Hansford & Chappell (1967). They have shown that Ca^{2+} stimulates α-glycerophosphate dehydrogenase. Moreover, the concentration of free Ca^{2+} that is needed is within the physiological range; half-maximal rates of oxidation are obtained with concentrations less than $0.1 \mu M$ (Fig. 5). Significantly, the concentration of Ca^{2+} necessary for activation of α-glycerophosphate dehydrogenase is very similar to that required to activate actomyosin ATPase (Chaplain, 1967) and phosphorylase *b* kinase (Hansford & Sacktor, 1970*a*) of flight muscle. As also shown in Fig. 5, the mechanism of action of Ca^{2+} in activating α-glycerophosphate dehydrogenase is to lower the apparent K_m for α-glycerophosphate (Hansford & Chappell, 1967). A plot of enzyme activity versus concentration of substrate is sigmoidal in the absence of Ca^{2+}. At 2mM-α-glycerophosphate, which is the concentration found in flight muscle (Sacktor & Wormser-Shavit, 1966), Ca^{2+} effects a 10-fold increase in rate of oxidation. This is greater than the threefold enhancement of the respiratory rate that is obtained

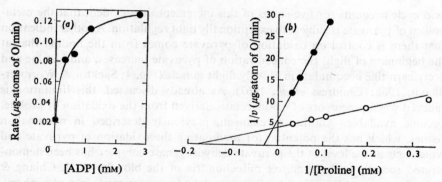

Fig. 6. *Control of proline oxidation by ADP with mitochondria isolated from flight muscle of the blowfly P. regina*

(*a*) The effect of the concentration of ADP on the rate of oxidation. The uncoupling agent carbonyl cyanide *p*-trifluoromethoxyphenylhydrazone is in the reaction mixture. (*b*) The kinetics of the oxidation of proline in the absence of ADP (●), and with 2.3 mM-ADP added (○). From Hansford & Sacktor (1970*b*).

by the addition of ADP to mitochondrial reactions oxidizing α-glycerophosphate in controlled (state 4) systems.

The concentration of proline in flight muscle of the blowfly is extraordinarily high, nearly 7 mM (Sacktor & Wormser-Shavit, 1966). This concentration decreases abruptly on initiation of flight and it is suggested that the mitochondrial oxidation of proline is facilitated by the rest-to-flight transition. As shown in Fig. 6, oxidation of proline by isolated blowfly flight-muscle mitochondria is activated by ADP in the presence of uncoupling agents as well as in the presence of oligomycin or atractyloside (Hansford & Sacktor, 1970*b*). Stimulation by ADP is also seen when sonically prepared submitochondrial particles are used and when phenazine methosulphate is used as the electron acceptor to by-pass the respiratory chain. These findings rule out the possibilities that the activation by the nucleotide is related to the penetration of proline into the mitochondrion and that ADP is acting at the level of the respiratory chain. Rather, these results indicate that the site of action of ADP is proline dehydrogenase itself. Also illustrated in Fig. 6 are experiments showing that ADP is an allosteric effector of the dehydrogenase and its mode of action is to lower the apparent K_m for proline (Hansford & Sacktor, 1970*b*). Significantly, the apparent K_m is lowered from 33 to 6 mM, the latter approximating the concentration of proline found in the muscle of the resting blowfly. The rate of proline oxidation is dependent on the relative proportions of the adenine nucleotides as well as on the absolute concentration of ADP (Hansford & Sacktor, 1970*b*). The rate is particularly sensitive to small increases in ADP in the presence of a high percentage of ATP. This suggests that the relatively small changes in concentrations of ATP and ADP in flight muscle, which occur when blowflies begin to fly (Sacktor & Hurlbut, 1966), may lead to appreciable increases in the rate of proline oxidation.

Since the respiratory rate of the flying insect may be increased 100 times that of the same insect at rest and the oxidation of pyruvate via the tricarboxylic acid cycle accounts for five-sixths of this increment, it is evident that the metabolism of pyruvate is subject to exceptionally tight regulation. Another indication that there is control on oxidation of pyruvate comes from the finding that at the beginning of flight the concentration of pyruvate increases, and alanine and acetylcarnitine accumulate in blowfly flight muscle (Fig. 4; Sacktor & Wormser-Shavit, 1966; Childress et al., 1967). As already discussed, this limitation is relieved when precursors of oxaloacetate, derived from the oxidation of proline, become available. Another mechanism, previously described in mammalian tissues, which has the potential for modulating the oxidation of pyruvate and which acts at the level of the pyruvate dehydrogenase complex has been demonstrated recently in flight-muscle mitochondria of the blowfly (P. K. Chiang & B. Sacktor, unpublished work). Pyruvate dehydrogenase exists in a phosphorylated or dephosphorylated form. Only the dephosphorylated form is catalytically active. In the presence of ATP, pyruvate dehydrogenase kinase converts the active enzyme into its inactive phosphorylated form. Of interest, the kinase is inhibited by pyruvate. Thus in situations in which the concentration of pyruvate rises, such as at the initiation of flight, its oxidation is facilitated. A specific phosphatase cleaves the protein–phosphate bond and converts the enzyme into an active form. Significantly, the phosphatase is Ca^{2+}-activated at concentrations of the bivalent cation found in the muscle. The physiological relevance of these enzymes and how their actions are correlated with other mechanisms (discussed below) for controlling the rate of pyruvate oxidation in flight muscle remains to be established.

The finding that the NAD^+-linked isocitrate dehydrogenase is subject to tight regulation has led to the suggestion that this enzyme may limit pyruvate oxidation in the tricarboxylate cycle in insect mitochondria. Goebell & Klingenberg (1964) have demonstrated that isocitrate dehydrogenase from locust flight-muscle mitochondria is markedly activated by ADP and isocitrate, and is inhibited by ATP. Later work with various insect species has extended these findings and has shown additionally that isocitrate is activated by citrate and H^+ as well as by ADP, phosphate and isocitrate, and is inhibited by NADH and Ca^{2+} in addition to ATP (Lennie & Birt, 1967; Hansford & Chappell, 1968; Vaughan & Newsholme, 1969; Ku & Cochran, 1971; Zahavi & Tahori, 1972; Hansford, 1972). The mode of action of ADP is to lower the apparent K_m for isocitrate. In the presence of a concentration of isocitrate approximating to that found in the mitochondrion, there is a 20-fold increase in isocitrate dehydrogenase activity on adding ADP (Hansford & Chappell, 1968). Since the stimulation by ADP and inhibition by ATP are dependent on the concentrations of these effectors, the activity of the dehydrogenase is largely determined by the relative proportion of the two nucleotides in a mixture of a fixed total concentration of adenine nucleotide. Indeed, Johnson & Hansford (1975) have shown that the ATP/ADP ratio is the main determinant of flux through the tricarboxylic acid cycle, with the redox state of nicotinamide nucleotide being of lesser importance.

In a series of elegant studies in our laboratories, Hansford (1974) and Johnson & Hansford (1975) have demonstrated convincingly that it is isocitrate dehydrogenase that primarily limits the oxidation of pyruvate. They have measured the concentration of many of the intermediates of the citric acid cycle

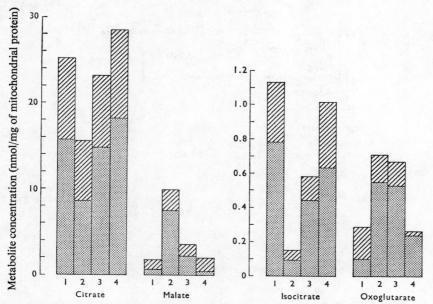

Fig. 7. *Changes in the concentration of tricarboxylate-cycle intermediates during the oxidation of pyruvate+ATP+bicarbonate in blowfly flight-muscle mitochondria during the transition from state 4 to state 3 to state 4*

Data obtained from Johnson & Hansford (1975). The mitochondrial intermediates are

$$\text{State 4 (1)} \xrightarrow{\text{ADP}} \text{State 3 (2 and 3)} \rightarrow \text{State 4 (4)}$$

▨, Mitochondrial; ▨▨, total.

both in isolated blowfly flight-muscle mitochondria during the state 4 to 3 transition and in flight muscle during the rest-to-flight transition. As illustrated in Fig. 7, they have found that the concentrations of metabolites preceding the isocitrate dehydrogenase reaction decrease on activation of oxidation, whereas the concentrations of intermediates subsequent to the dehydrogenase increase in concentration during the transition. This cross-over at the isocitrate dehydrogenase reaction, concomitant with an increased flux through the system, clearly identifies the point of control.

Perhaps of additional importance to the mechanism of regulation of isocitrate dehydrogenase and the control of pyruvate oxidation is the finding of Vaughan & Newsholme (1969) that the effect of ADP on the dehydrogenase is dependent on the concentration of Ca^{2+}. At a minimal Ca^{2+} concentration, approximately 1 nM, the enzyme is maximally active (at a given isocitrate concentration) in the absence of ADP. At $10\,\mu M$-Ca^{2+}, however, in the absence of added ADP, activity is extremely low but is increased by the addition of ADP. Both Ca^{2+} and ADP affect the apparent K_m of the dehydrogenase for isocitrate. However, raising the concentration of ADP decreases the apparent K_m whereas raising the concentration of Ca^{2+} increases the apparent K_m. The effects of Ca^{2+} and ADP on the enzyme are independent. This action of Ca^{2+} on isocitrate dehydrogenase takes on added significance with the recent findings of B. Bulos, B. J. Thomas & B. Sacktor (unpublished work) that Ca^{2+} at low physiological concentrations

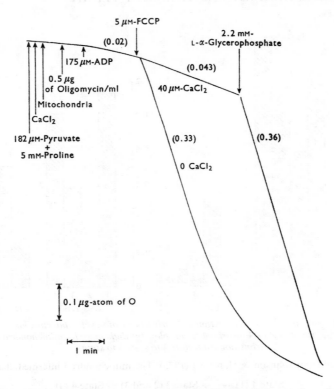

Fig. 8. *Inhibition of the oxidation of pyruvate by blowfly flight-muscle mitochondria by physiological concentrations of Ca²⁺*

Additions to the reaction mixture are indicated by arrows. Numbers along each line show rates of respiration expressed as μg-atoms of O/min per mg of mitochondrial protein. (B. Bulos, B. J. Thomas & B. Sacktor, unpublished work.) FCCP, carbonyl cyanide *p*-trifluoromethoxyphenyl-hydrazone.

specifically inhibits the oxidation of pyruvate by mitochondria isolated from the blowfly (Fig. 8).

The significant role of Ca^{2+} in regulating the metabolism of insect flight muscle has been emphasized. Therefore the capacity of mitochondria to translocate Ca^{2+} may be of utmost importance. This is even more crucial in the asynchronous flight muscle, wherein the sarcoplasmic reticulum is very sparse, and there is a major question as to the identity of the organelles participating in Ca^{2+} segregation and release. Carafoli *et al.* (1971) have found that Ca^{2+} is accumulated by respiring blowfly mitochondria. The uptake is supported by the oxidation of α-glycerophosphate and pyruvate plus proline, requires the presence of a permeant anion, and is prevented by respiratory-chain inhibitors, uncouplers, Ruthenium Red (Carafoli & Sacktor, 1972), but not by oligomycin. Although some information is available as to the mechanism by which Ca^{2+} is taken up by insect mitochondria, the mechanism by which intramitochondrial Ca^{2+} is released from the mitochondria is unknown. However, whatever the mechanism, it is apparent that the movements of Ca^{2+} into and out of blowfly flight-muscle mitochondria may be too slow to play a primary role in the rapid single contraction–relaxation

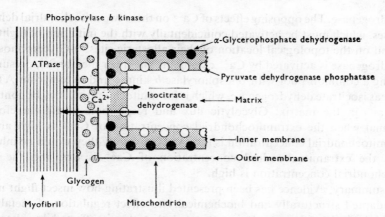

Fig. 9. *Diagrammatic illustration of the molecular orientation of the five enzymes that regulate flight-muscle metabolism whose activities are sensitive to physiological concentrations of Ca²⁺*

Glycolytic flux and respiration would be maximal when the extramitochondrial concentration of Ca^{2+} is high and the intramitochondrial concentration is low, and conversely they would be inhibited when the extramitochondrial concentration of Ca^{2+} is low and the intramitochondrial concentration is high. From Sacktor (1975).

cycles of this muscle during flight. Moreover, since blowfly muscle has a much decreased sarcoplasmic reticulum, it seems that segregation of Ca^{2+} within these membrane vesicles is not a prerequisite for the single contraction–relaxation cycle of this asynchronous muscle. On the other hand, Carafoli *et al.* (1971) have conjectured that it is possible for Ca^{2+} to accumulate slowly during a train of successive contraction–relaxation cycles and then to be released when the next nerve impulse arrives.

In the present discussion, it has been pointed out that at least five enzymes participating in the regulation of the rest-to-flight transition are now known to be sensitive to the Ca^{2+} at the level of the bivalent cation present in blowfly muscle, 800 nmol/g wet wt. (Carafoli *et al.*, 1971). As illustrated diagrammatically in Fig. 9, two of these, actomyosin ATPase (Sacktor, 1953; Jewell & Ruegg, 1966) and phosphorylase *b* kinase (Hansford & Sacktor, 1970*a*; Sacktor *et al.*, 1974), are extramitochondrial and both are activated by Ca^{2+}. The other three Ca^{2+}-dependent enzymes are mitochondrial; namely, α-glycerophosphate dehydrogenase (Estabrook & Sacktor, 1958; Hansford & Chappell, 1967), pyruvate dehydrogenase phosphatase (P. K. Chiang & B. Sacktor, unpublished work) and NAD⁺-linked isocitrate dehydrogenase (Vaughan & Newsholme, 1969). As shown in Fig. 9, the mitochondrial α-glycerophosphate dehydrogenase is located on the outer surface of the inner membrane (Klingenberg, 1970) and the NAD⁺-linked isocitrate dehydrogenase is present in the mitochondrial matrix (Goebell & Klingenberg, 1963). The location of the pyruvate dehydrogenase phosphatase is not known at present, but it is tentatively placed at the inner surface of the cristal membrane. It is important to note, however, that for the two dehydrogenases, whose locations are known, the control by Ca^{2+} is different. Ca^{2+} is an allosteric activator of α-glycerophosphate dehydrogenase, whereas the bivalent cation is an allosteric inhibitor of isocitrate dehydrogenase. The opposing effects of Ca^{2+} on these two mitochondrial dehydrogenases, which must be activated coincidentally with the initiation of flight, may depend on the topological location of their allosteric sites. α-Glycerophosphate dehydrogenase is activated by Ca^{2+} external to the inner membrane, presumably sharing a common pool with phosphorylase b kinase and actomyosin ATPase, whereas isocitrate dehydrogenase, which is intramitochondrial, may be controlled by Ca^{2+} in the matrix. Glycolytic flux and respiration would therefore be maximal when the extramitochondrial concentration of Ca^{2+} is high and the intramitochondrial concentration is low, and conversely they would be inhibited when the extramitochondrial concentration of Ca^{2+} is low and the intramitochondrial concentration is high.

In summary, evidence has been presented illustrating how insect flight muscle has adapted structurally and biochemically to effect regulation of metabolism and biological oxidations. It is suggested that this is effected by alterations in the flux of Ca^{2+} and the phosphate potential, i.e. the relationship between the concentration of ATP relative to those of ADP, AMP and phosphate, and accomplished via co-ordination of concerted actions of these effectors at specific loci.

References

Azzone, G. F. & Carafoli, E. (1960) *Exp. Cell Res.* **21**, 447–467
Brosemer, R. W. (1967) *J. Insect Physiol.* **13**, 685–690
Bücher, T. & Klingenberg, M. (1958) *Angew. Chem. Int. Ed. Engl.* **70**, 552–570
Bulos, B., Shukla, S. & Sacktor, B. (1972) *Arch. Biochem. Biophys.* **149**, 461–469
Bursell, E. (1963) *J. Insect Physiol.* **9**, 439–452
Bursell, E. (1965) *Comp. Biochem. Physiol.* **16**, 259–266
Bursell, E. (1966) *Comp. Biochem. Physiol.* **19**, 809–818
Bursell, E. (1967) *Comp. Biochem. Physiol.* **23**, 825–829
Carafoli, E. & Sacktor, B. (1972) *Biochem. Biophys. Res. Commun.* **49**, 1498–1503
Carafoli, E., Hansford, R. G., Sacktor, B. & Lehninger, A. L. (1971) *J. Biol. Chem.* **246**, 964–972
Chance, B. & Sacktor, B. (1958) *Arch. Biochem. Biophys.* **76**, 509–531
Chaplain, R. A. (1967) *Biochim. Biophys. Acta* **131**, 385–392
Chappell, J. B. & Crofts, A. R. (1966) in *Regulation of Metabolic Processes in Mitochondria* (Tager, J. M., Papa, S., Quagliariello, E. & Slater, E. C., eds.), pp. 293–316, Elsevier, Amsterdam
Chefurka, W. (1958) *Biochim. Biophys. Acta* **28**, 660–661
Childress, C. C. & Sacktor, B. (1966) *Science* **154**, 268–270
Childress, C. C. & Sacktor, B. (1970) *J. Biol. Chem.* **245**, 2927–2936
Childress, C. C., Sacktor, B. & Traynor, D. R. (1967) *J. Biol. Chem.* **242**, 754–760
Childress, C. C., Sacktor, B., Grossman, I. W. & Bueding, E. (1970) *J. Cell Biol.* **45**, 83–90
Clark, J. B. & Nicklas, W. J. (1970) *J. Biol. Chem.* **245**, 4724–4731
Crabtree, B. & Newsholme, E. A. (1970) *Biochem. J.* **117**, 1019–1021
Crabtree, B. & Newsholme, E. A. (1972) *Biochem. J.* **126**, 49–58
Davis, R. A. & Fraenkel, G. (1940) *J. Exp. Biol.* **17**, 402–407
deKort, C. A. D., Bartelink, A. K. M. & Schuurmans, R. R. (1973) *Insect Biochem.* **3**, 11–17
Delbruck, A., Zebe, E. & Bücher, T. (1959) *Biochem. Z.* **331**, 273–296
Estabrook, R. W. & Sacktor, B. (1958) *J. Biol. Chem.* **233**, 1014–1019
Goebell, H. & Klingenberg, M. (1963) *Biochem. Biophys. Res. Commun.* **13**, 209–212
Goebell, H. & Klingenberg, M. (1964) *Biochem. Z.* **340**, 441–464
Hansford, R. G. (1972) *Biochem. J.* **127**, 271–283
Hansford, R. G. (1974) *Biochem. J.* **142**, 509–519
Hansford, R. G. & Chappell, J. B. (1967) *Biochem. Biophys. Res. Commun.* **27**, 686–692
Hansford, R. G. & Chappell, J. B. (1968) *Biochem. Biophys. Res. Commun.* **30**, 643–648
Hansford, R. G. & Johnson, R. N. (1975) *Biochem. J.* **148**, 389–401

Hansford, R. G. & Sacktor, B. (1970a) *FEBS Lett.* **7**, 183–187
Hansford, R. G. & Sacktor, B. (1970b) *J. Biol. Chem.* **245**, 991–994
Hansford, R. G. & Sacktor, B. (1971) in *Chemical Zoology* (Florkin, M. & Scheer, B. T., eds.),
   vol. 6, pp. 213–247, Academic Press, New York
Hedman, R. (1965) *Exp. Cell Res.* **38**, 1–12
Jewell, B. R. & Ruegg, J. C. (1966) *Proc. R. Soc. London Ser. B.* **164**, 428–459
Johnson, R. N. & Hansford, R. G. (1975) *Biochem. J.* **146**, 527–535
Kirsten, E., Kirsten, R. & Arese, P. (1963) *Biochem. Z.* **337**, 167–178
Kitto, G. B. & Briggs, M. H. (1962a) *Nature (London)* **193**, 1003–1004
Kitto, G. B. & Briggs, M. H. (1962b) *Science* **135**, 918
Klingenberg, M. (1970) *Eur. J. Biochem.* **13**, 247–252
Klingenberg, M. & Bücher, T. (1960) *Annu. Rev. Biochem.* **29**, 669–698
Klingenberg, M. & Slenczka, W. (1959) *Biochem. Z.* **331**, 334–336
Ku, T. Y. & Cochran, D. G. (1971) *Insect Biochem.* **1**, 81–96
Lennie, R. W. & Birt, L. M. (1967) *Biochem. J.* **102**, 338–350
Maruyama, K., Pringle, J. W. S. & Tregear, R. T. (1968) *Proc. R. Soc. London Ser. B* **169**, 229–240
Norden, D. A. & Venturas, D. J. (1972) *Insect Biochem.* **2**, 226–234
Pette, D., Luh, W. & Bücher, T. (1962) *Biochem. Biophys. Res. Commun.* **7**, 419–424
Pringle, J. W. S. & Tregear, R. T. (1969) *Proc. R. Soc. London Ser. B* **174**, 33–50
Reed, W. D. & Sacktor, B. (1971) *Arch. Biochem. Biophys.* **145**, 392–401
Sacktor, B. (1953) *J. Gen. Physiol.* **36**, 371–387
Sacktor, B. (1955) *J. Biophys. Biochem. Cytol.* **1**, 29–46
Sacktor, B. (1965) in *The Physiology of Insecta* (Rockstein, M., ed.), vol. 2, pp. 483–580,
   Academic Press, New York
Sacktor, B. (1970) *Adv. Insect Physiol.* **7**, 267–348
Sacktor, B. (1975) in *Insect Biochemistry and Function* (Candy, D. J. & Kilby, B. A., eds.),
   pp. 1–88, Chapman and Hall, London
Sacktor, B. & Childress, C. C. (1967) *Arch. Biochem. Biophys.* **120**, 583–588
Sacktor, B. & Dick, A. (1962) *J. Biol. Chem.* **237**, 3259–3263
Sacktor, B. & Hurlbut, E. C. (1966) *J. Biol. Chem.* **241**, 632–634
Sacktor, B. & Wormser-Shavit, E. (1966) *J. Biol. Chem.* **241**, 624–631
Sacktor, B., Wu, N-C., Lescure, O. & Reed, W. D. (1974) *Biochem. J.* **137**, 535–542
Slack, E. N. (1975) Ph.D. Thesis, University of London
Tulp, A. & Van Dam, K. (1970) *FEBS Lett.* **10**, 292–294
Tyler, D. D. & Gonze, J. (1967) *Methods Enzymol.* **10**, 75–77
Van den Bergh, S. G. (1967) in *Mitochondrial Structure and Compartmentation* (Quagliariello,
   E., Papa, S., Slater, E. C. & Tager, J. M., eds.), pp. 203–206, Adriatica Editrice, Bari
Van den Bergh, S. G. & Slater, E. C. (1962) *Biochem. J.* **82**, 362–371
Vaughan, H. & Newsholme, E. A. (1969) *FEBS Lett.* **5**, 124–126
Weis-Fogh, T. (1952) *Phil. Trans. R. Soc. London Ser. B* **237**, 1–36
Weis-Fogh, T. (1964) *J. Exp. Biol.* **41**, 229–256
Weis-Fogh, T. (1967) *J. Exp. Biol.* **47**, 561–587
Zahavi, M. & Tahori, A. S. (1972) *J. Insect Physiol.* **18**, 609–614
Zebe, E. C. & McShan, W. H. (1957) *J. Gen. Physiol.* **40**, 779–790

Biochem. Soc. Symp. (1976) **41**, 133–168
Printed in Great Britain

Facultative Anaerobiosis in Molluscs

By A. DE ZWAAN, J. H. F. M. KLUYTMANS and D. I. ZANDEE

*Laboratory of Chemical Animal Physiology, State University of Utrecht
Transitorium III, 8 Padualaan, Utrecht, The Netherlands*

Synopsis

The glycolytic fermentation of molluscs is rather complex. Multiple end products accumulate (lactate, alanine, octopine, succinate, propionate, acetate and CO_2), which are partly formed in the cytoplasm and partly in the mitochondrion. Various schemes have been presented to account for these end products as well as for the maintenance of the redox balance. With respect to the role of alanine there are two opinions: (1) alanine accumulation is continuous and is essential for the generation of the mitochondrial NADH required in the reduction of fumarate and (2) succinate and alanine (initial end products) accumulate in different compartments and their accumulation occurs independently. Both statements are evaluated in the light of the latest experimental observations including the regulatory properties at the phosphoenolpyruvate branchpoint and the effect of pH and 'energy charge'. For nervous tissue the function of oxygen can be replaced by the lipochrome pigment, which enables carbohydrates to be totally oxidized to CO_2 and water. The simultaneous mobilization of carbohydrates and amino acids is not supported by the experimental data. Various advantages of the glycolytic fermentation in molluscs as compared with classical glycolysis in skeletal muscle are discussed.

Introduction

In contrast with the situation in terrestrial organisms facultative anaerobiosis is a common feature among aquatic organisms, especially the invertebrates [for a review see von Brand (1946)]. Facultative anaerobes use oxygen when available, but can rely on fermentation for some time during its absence. Molluscs are very suitable for studies on the relationship between energy metabolism and oxygen availability, because some species are complete aerobes whereas others are facultative anaerobes. Although obligate-anaerobic species do not occur, the manner of ATP generation of facultative-anaerobic molluscs during anaerobiosis shows great similarities to that found in obligate-anaerobic invertebrates, such as parasitic worms. Among the molluscs, bivalves generally show the best developed adaptations to anaerobiosis, gastropods show less adaptation, whereas cephalopods cannot survive this condition. This can be related to their habitats, which for many bivalves show great fluctuations in oxygen concentrations, e.g. bivalves living in the intertidal zone, or in or on the bottom layer of shallow waters. Bivalves show the least locomotion, especially those species with the best-developed anaerobic functions (for example the sessile bivalves of the intertidal zone show high anoxic tolerance, whereas the active swimming Pectinidae are almost strict aerobes) and store the highest amounts of glycogen of all molluscs (Giese, 1969). Bivalves together with opistobranch gastropods lack haemocyanin, the main respiratory pigment in molluscs, whereas the blood of cephalopods contains high concentrations (Ghiretti & Ghiretti-Magaldi, 1972). The relationship between habitat and anoxic tolerance in bivalves makes this group a favourite in studies on metabolic adaptations to anaerobiosis and this explains why most references cited in this review deal with them.

ATP Generation

One of the elementary properties of life is a continuous generation of ATP to enable energy-requiring functions such as transport, mechanical work and biosynthetic processes to operate. ATP synthesis is coupled with the enzymic degradation of nutrient molecules (carbohydrates, lipids or proteins). The role of carbohydrates (glycogen, trehalose, glucose), lipids and proteins as energy stores in bivalve molluscs has been reviewed by de Zwaan & Wijsman (1975). During aerobiosis there are preferences for certain substances depending on the season. Carbohydrates form the main energy reserves under anaerobic conditions, but fatty acid and amino acid conversion cannot be ruled out.

The degradation of energy substrates can lead to ATP synthesis by two methods: (1) the synthesis of energy-righ phosphate esters or other intermediates which have high 'group potentials' (e.g. succinyl-CoA) from which ATP can be obtained and (2) the oxidation of NADH (and $FADH_2$) via the electron-transfer chain. Which system will dominate depends on the type of electron acceptors available in a specific situation. In the absence of oxygen, when the final electron acceptor is an organic molecule, which is usually generated in the fermentation process itself, the first method (the substrate-level phosphorylation) is responsible for ATP synthesis. In respiration, oxygen is the final electron acceptor, which enables phosphorylation to be coupled to the electron-transport chain by the second method. At the same time substrate-level phosphorylations are also performed. Respiration has great advantages over fermentation with respect to the ATP yield; glucose degradation by respiration yields 18 times more ATP than by fermentation to lactate.

Until recently it was generally assumed that in obligate anaerobes and in facultative anaerobes during anoxia, ATP synthesis depends entirely on substrate-level phosphorylations. However, recent evidence has shown that electron transfer coupled to ATP synthesis can take place in the absence of oxygen in facultative anaerobes and even in organisms that do not use oxygen at all. Studies by Stadtman & Elliot (1956) and Stadtman et al. (1958) on anaerobic micro-organisms belonging to the genus Clostridium revealed that the anaerobic reductive deamination of glycine to acetate and NH_3 is coupled with the formation of ATP by a mechanism independent of the formation of substrate-level high-energy intermediates. Indirect evidence for the potential importance of anaerobic electron-transport-coupled phosphorylation in the energy metabolism of anaerobic bacteria has been discussed by Stadtman (1966). For the liver fluke *Fasciola hepatica* it has been shown that NADH derived from glycolysis is reoxidized by a 'fumarate–oxidoreductase enzyme complex' in such a way that the hydrogen is transferred via a flavoprotein to quinone and subsequently via a second flavoprotein to fumarate, giving rise to succinate formation. Furthermore, it was found that the oxidation of NADH was coupled with phosphorylation in a ratio of approximately 1:1 (de Zoeten & Tipker, 1969). Oxidation of NADH by fumarate coupled with the synthesis of ATP was also found to occur in cyanide-poisoned rat heart submitochondrial particles. Cytochrome *b* was oxidized in this reaction, and amytal was an inhibitor. These facts also suggest that a portion of the electron-transfer chain, including cytochrome *b*, is involved (Wilson & Cascarano, 1970). The reaction also occurs in cytochrome oxidase-deficient sarcosomes of *Ascaris lumbricoides* (Seidman & Entner, 1961). An NADH-dependent fumarate reduction has also been shown in various bivalves (Hammen & Lum, 1966; Wegener *et al.*, 1969; Chen & Awapara, 1969; Hammen, 1969; O'Doherty & Feltham, 1971; Hammen, 1975). Singer (1971) observed that the ratio of fumarate reductase (fumarate→succinate) to succinate dehydrogenase (succinate→fumarate) increases as one goes from strict aerobes through facultative anaerobes to obligate anaerobes. Bivalve molluscs exhibit varying ratios; in six species the ratios ranged from 0.2 to 7 (Hammen & Lum, 1966). It appears that a high ratio reflects an adaptation to anaerobic conditions, because the ratio corresponds to the degree of average oxygen availability of the habitat.

Direct evidence is not available to show that the NADH-dependent fumarate reduction in bivalves is coupled with a phosphorylation. An attempt to prove this connexion in homogenates of *Crassostrea virginica* was unsuccessful (Wegener *et al.*, 1969). Hammen (1975) reports an acceleration of fumarate reduction by ADP and together with the observed absolute dependence on ATP of the reverse reaction (NAD^+-linked succinate oxidation) these facts provide strong evidence for a coupled phosphorylation accompanying fumarate reduction. The apparent K_m value, however, was very high for fumarate, ranging from about 25 to 30mM in the three bivalves studied. Moreover, the evidence presented for ADP stimulation is equivocal, because a control containing both ADP and fumarate but without NADH was lacking (see Fig. 2 from Hammen, 1975).

Another system in molluscs in which a part of the electron-transfer chain is involved, is 'anoxic endogenous oxidation' (see under 'Oxygen is replaced by a lipochrome pigment'). In this case the final electron acceptor is not formed from the energy substrates (as is true for fumarate), but is stored in particles called cytosomes as a pigment composed of lipids, the lipochrome.

It has been deduced from various observations that life arose under anaerobic conditions. Glycolytic fermentations in which fuel substrates are converted into intermediates with high 'group potentials' from which the potential energy can be used for biosynthetic processes might have played a dominant role in early stages of biological evolution. Gradually these systems might have evolved in such a way that electron-transfer-coupled ATP generation became an integral part of the overall pathway, but with fumarate as the final electron acceptor. Gradually after the evolution of photosynthetic organisms, oxygen became available. Life expanded from the sea to the land. From the fact that the present atmosphere contains about 20 times more oxygen per litre than does the hydrosphere, it will be clear that oxygen is more available in terrestrial habitats. It is therefore understandable that the organisms that migrated to land evolved an energy metabolism which relied primarily on respiration. The system of electron-transfer-coupled ATP synthesis would become more efficient by replacing fumarate by oxygen. To make optimal use of this system it is necessary that the fuel molecule is metabolized by a reaction sequence in which it is completely oxidized. Then the maximal amount of NADH becomes available for the electron-transport chain. This happens when intermediates are formed which can be channelled into the tricarboxylic acid cycle running in the forward direction. Besides one oxidation, the route glucose→fumarate involves two reduction steps (see under 'Oxygen is replaced by fumarate'), and therefore might have become rudimentary. This may explain why in mammalian glycolytic tissue [bovine heart, rat heart, rat gastrocnemius muscle, see Sanadi & Fluharty (1963), Haas (1964) and Wilson & Cascarano (1970)] reduction of fumarate by NADH has been shown in experiments *in vitro*, but the enzyme phosphoenolpyruvate-carboxykinase, which is of vital importance in the formation of fumarate (see under 'Fumarate formation'), is more or less absent from these tissues (see Scrutton & Utter, 1968).

Glycolytic Fermentation in Molluscs Compared with Skeletal Muscle of Mammals

The glycolytic fermentation of skeletal muscle (Embden–Meyerhof–Parnas route) is relatively less complex than in molluscs, as only one class of fermentable substrate (carbohydrates) is involved, which is converted into the sole end product lactate. The sequence of reactions involves one oxidation, catalysed by glyceraldehyde 3-phosphate dehydrogenase, and one reduction, catalysed by lactate dehydrogenase. The 1:1 organization of the glyceraldehyde 3-phosphate dehydrogenase/lactate dehydrogenase redox reactions maintains a constant redox state in the cell. Both the hydrogen donor (glyceraldehyde 3-phosphate) and hydrogen acceptor (pyruvate) are formed in the same compartment of the cell, the cytoplasm. The intermediates of the coupled redox reactions are separated in the reaction sequence, and therefore the coenzyme NAD mediates the hydrogen transport.

During anoxia facultative anaerobic molluscs accumulate a variety of organic molecules such as succinate, propionate, acetate, alanine and CO_2 in addition to small amounts of lactate (Rosen, 1966; de Zwaan *et al.*, 1975a; Kluytmans *et al.*, 1975; Gäde, 1975; Gäde *et al.*, 1975; Gäde & Grieshaber, 1975; de Zwaan *et al.*, 1976; Kluytmans & de Zwaan, 1976). Glycolytic fermentation in molluscs therefore consists of many coupled redox reactions. Moreover, some of the end products are formed in the cytoplasm (octopine, lactate and probably alanine) and others in the mitochondrion (acetate, succinate, propionate and CO_2). The inner mitochondrial membrane is impermeable to NAD^+ and therefore a barrier has been introduced against the coenzyme that couplex redox reactions in the reaction sequence. This implies that the redox balance must be maintained on both sides of the barrier or that shuttle systems must be involved.

The ratio in which the various intermediates and end products accumulate alters with the length of time of anaerobiosis. Alanine and lactate are probably initial end products. Species differences, however, have been shown; for example acetate accumulates to a greater degree in *Anodonta cygnea* than in *Mytilus edulis* [for a detailed discussion see de Zwaan & Wijsman (1975)]. It is therefore understandable why it is not possible to give one general fermentation scheme for molluscs, representing all species and showing all stages of anaerobiosis.

Since the paper of Stokes & Awapara (1968), a number of different schemes have been described. Despite their differences, all the schemes for carbohydrate fermentation have one feature in common, namely the sequence of reactions leading to succinate accumulation (see under 'Schemes of glycolytic fermentation'). This route proceeds via classical glycolysis (Embden–Meyerhof–Parnas scheme) up to the stage of phosphoenolpyruvate (or pyruvate); this is then carboxylated to oxaloacetate. Oxaloacetate is then converted into succinate via part of the citric acid cycle operating in the reverse direction.

Metabolic Responses to Anaerobiosis

Use of stored oxygen

When intertidal bivalves close their shells, some seawater is trapped in the mantle cavity and this can serve as a source of oxygen. In *M. edulis* about 3 ml of water, containing no more than $30 \mu l$ of oxygen, becomes enclosed. This is equivalent to the production of

$$\frac{30}{22.4} \times 2 \times \frac{14}{5} = 7.5 \mu\text{mol of ATP}$$

(14/5 is the average ATP yield per O atom in the tricarboxylic acid cycle). The energy demand of *M. edulis* during anaerobiosis is about $15 \mu\text{mol}$ of ATP/h at room temperature [de Zwaan & Wijsman (1975); average over the first 48 h of anoxia]. Therefore this source of oxygen can only meet the energy demand for about 30 min. Measurements of oxygen tension in the extrapallial fluid of *Mercenaria mercenaria* by Crenshaw & Neff (1969) revealed that the clam becomes completely anaerobic within 25 min of shell closure.

Some oxygen can be stored in the tissues, especially as myoglobin is present in adductor muscle and heart (Ghiretti & Ghiretti-Magaldi, 1972). The same might be true for the blood, which in some species contains haemoglobin, but this source can only contribute to a short initial period of anaerobiosis (Zs.-Nagy, 1973).

Karnaukhov (1971) assumed that the carotenoids present in the cytosomes (see also under 'Oxygen is replaced by lipochrome pigment') of the nervous system could bind oxygen by means of their conjugated bonds. Ganglion tissue of *A. cygnea* contains about 10 mg of carotenoid per 100 g wet weight (Lábos *et al.*, 1966). As calculated by Zs.-Nagy (1974) this might fix $90 \mu\text{mol}$ of oxygen if each carotene molecule is completely saturated. The oxygen consumption of 100 g (wet wt.) of ganglia is at least 940 μmol/h (Zs.-Nagy, 1971b) so that the oxygen stored on carotenoids could therefore meet the demand for only about 6 min at the normal metabolic rate. However, if the metabolic rate of nervous tissue falls to that of the body tissue, for *M. edulis* the ATP demand is decreased to about 5% (de Zwaan & Wijsman, 1975), the demand could be met for a few hours.

Summarizing, we can state that after shell closure there is no immediate conversion of the mollusc to anaerobic conditions.

Oxygen is replaced by a lipochrome pigment

The nervous tissue, and other tissues, of molluscs contain pigmented cell particles, called cytosomes. The yellow pigment, lipochrome, is composed of neutral fats and phospholipids (Zs.-Nagy & Csukàs, 1969). Cytosomes are particularly found in animals that can survive anoxic conditions for long periods (Zs.-Nagy, 1971a). They contain mitochondrial enzymes (succinate dehydrogenase, cytochrome oxidase), and morphogenic studies also support a mitochondrial origin for them (Zs-Nagy, 1969). Zs.-Nagy postulated that cytosomes are a kind of anaerobic mitochondria in which the liposome pigment serves as a temporary electron acceptor in the electron-transfer chain [for a review see Zs.-Nagy (1974)].

The assumption that the cytosomes take part in the anaerobic oxidative-energy production of the cell is supported by the fact that they accumulate bivalent cations from the medium, even in the absence of oxygen. This process is inhibited by dinitrophenol and CN^-, indicating the participation of cytochrome oxidase and coupling mechanisms in anoxic energy production. In this way 'anoxic endogenous oxidation' is able to realize about 60% of the total energy production from carbohydrate molecules without environmental oxygen (Zs.-Nagy, 1973).

Zs.-Nagy assumes that the type of energy-yielding mechanism during anaerobiosis is carbohydrate metabolism totally or partly via 'anoxic endogenous oxidation' (Zs.-Nagy & Ermini, 1972; Zs.-Nagy, 1973), which implies that glycolytic fermentation would be of minor importance. He advances the following arguments in favour of this assumption.

(1) In *Mytilus galloprovincialis* ATP concentrations during anoxia reached between 57.1 and 60.9% of the normal values (Zs.-Nagy & Ermini, 1972). In *A. cygnea* 52–94% of the normal ATP concentrations were detected during 7 days of anoxia, while the carbohydrate consumption remained unchanged (Zs.-Nagy, 1973). The agreement with the calculated efficiency of about 60% compared with oxygen as electron acceptor would be conclusive in favour of the theory of anoxic endogenous oxidation.

(2) Biochemical measurements have failed to show any agreement between the amount of carbohydrates consumed and the organic acids formed during anoxia (Zs.-Nagy, 1971a). This indicates that at least part of the carbohydrate becomes degraded by some other mechanisms.

(3) If glycolytic fermentation is of major importance carbohydrate consumption must then be increased to several times the normal, which is certainly not the case.

Lactate production would also increase, but many studies have shown that this does not occur (Zs.-Nagy, 1973).

All these facts are certainly true [for detailed information see review by de Zwaan & Wijsman (1975)], but are not arguments in favour of the anoxic endogenous oxidation concept. Zs.-Nagy (1973) starts by assuming that ATP concentrations in an organism are equal to the production of this compound and he therefore believes that the energy requirement during anoxia is hardly decreased. That this is certainly not the case has been shown by de Zwaan & Wijsman (1975). Therefore there is no reason to expect an increase in carbohydrate consumption during anaerobiosis (Pasteur effect). With regard to the discrepancy in stoicheiometric studies, it is relevant to note that in bivalves volatile fatty acids are a main anaerobic end product of carbohydrate degradation (Kluytmans et al., 1975; Gäde et al., 1975). In the papers cited by Zs.-Nagy (1973), not all of the end products were determined.

If, as stated by Zs.-Nagy (1973), about the same amount of carbohydrate is consumed via the lipochrome system normally and during anaerobiosis, the implication is that CO_2 production from glucose should also be about the same during aerobic and anaerobic metabolism. The tricarboxylic acid cycle would then operate at the same speed under both conditions. However, experiments with L-[U-^{14}C]glucose and L-[U-^{14}C]glutamate in *M. edulis* show that the tricarboxylic acid cycle becomes less active during anaerobiosis (percentage ^{14}C in CO_2 is decreased from 22.9 to 2.0 and from 35.2 to 6.1 for glucose and glutamate respectively) (de Zwaan et al., 1975a).

Oxygen is replaced by fumarate

Fumarate formation. Hammen & Wilbur (1959) incubated mantle tissue of *C. virginica* in aerated seawater with ^{14}C-labelled bicarbonate or sodium [1-^{14}C]propionate. Both precursors were rapidly assimilated. Most radioactivity was found in intermediates of the tricarboxylic acid cycle and some in the amino acid, aspartate. In the organic acid fraction more than 90% of the activity was found in succinate, the remainder in fumarate and malate. The distribution of radioactive label was similar with both precursors and this was explained by assuming that CO_2 fixation started from propionate, which was carboxylated to succinate. More extensive studies (Hammen & Osborne, 1959; Hammen & Lum, 1962) showed that incorporation of CO_2 into citric acid-cycle intermediates is a common feature of invertebrates. Hammen (1964) also established for *Mer. mercenaria* that the radioactivity of $NaH^{14}CO_3$ is mainly fixed into succinate. In the same year two papers by Awapara and co-workers showed that CO_2 is not incorporated into propionate by the action of propionyl-CoA carboxylase, but into phosphoenolpyruvate by phosphoenolpyruvate carboxykinase. Awapara & Campbell (1964) studied the incorporation of CO_2 into amino acids in the oyster *C. virginica* and the brackish water clam *Rangia cuneata*. Labelled carbon was effectively incorporated into alanine, aspartate and glutamate. The incorporation proceeded at a slightly decreased rate in the presence of malonate at a concentration which is known to completely inhibit succinate oxidation. The results suggest carboxylation of phosphoenolpyruvate or pyruvate. Surprisingly, there was no incorporation into amino acids under anaerobic conditions. To obtain more evidence for this CO_2-fixation route, Simpson & Awapara (1964) checked other invertebrates, including six bivalves, for the presence of enzymes that catalyse carboxylation reactions. All bivalves showed high phosphoenolpyruvate carboxykinase activities compared with rat and chicken liver. Fixation of CO_2 into non-volatile acids in oyster mantle was increased significantly when phosphoenolpyruvate and ATP were added (Awapara & Campbell, 1964), which also indicates the presence of phosphoenolpyruvate carboxykinase. Other enzymes tested for were 'malic' enzyme, in the direction of NADPH oxidation, propionyl-CoA carboxylase and pyruvate carboxylase. Only 'malic' enzyme was found and only in *R. cuneata*. Awapara & Campbell (1964) concluded that in molluscs, phosphoenolpyruvate is probably the major source of the four-carbon dicarboxylic acids terminating in succinate.

Hammen (1966) studied the route of CO_2 fixation in mantle tissue of *C. virginica* by incubating tissue with $NaH^{14}CO_3$ for periods as brief as 3.6 s. Six tricarboxylic acid-cycle intermediates were labelled. After periods of less than 2 min most of the label was found in malate. Malate was also the first metabolite to show a continuous decline in radioactivity. Hammen (1966) therefore considered it to be the initial product of CO_2 fixation. This was further supported by the fact that tissue homogenates contained an $NADP^+$-linked malate dehydrogenase ('malic' enzyme), which was contrary to the result of Simpson & Awapara (1964). The reaction was assayed in the direction of $NADP^+$ reduction, the reverse reaction occurring only to a slight extent. The oyster also showed the highest ratio for pyruvate kinase/phosphoenolpyruvate carboxykinase in the study of Simpson & Awapara (1966) and this would favour the continuation of glycolysis to the stage of pyruvate (Hammen, 1969).

Simpson & Awapara (1966) presented a scheme for the formation of succinate from glucose. Phosphoenolpyruvate is formed from glucose by classical glycolysis and is converted by phosphoenolpyruvate carboxykinase into oxaloacetate, which in turn is converted into succinate via malate and fumarate. A rapid conversion of oxaloacetate into succinate was concluded from the accumulation of $[^{14}C]$succinate from $[^{14}C]$glucose in tissues of three different bivalves. The activities of malate dehydrogenase in these bivalves were high and in the same order as those found in the livers of chicken and rat. By studying the formation of succinate from $[6-^{14}C]$glucose and $[3-^{14}C]$pyruvate in the mantle of *R. cuneata*, the authors concluded that phosphoenolpyruvate reacts with CO_2 to form oxaloacetate and ultimately succinate. They found that approximately ten times more radioactivity was incorporated from glucose than from pyruvate. However, it is not known if there are permeability differences for the two precursors. A difficult point to understand in their concept is the high activities of pyruvate kinase found in bivalve tissues. These activities are as high as, or mostly considerably higher than, those of phosphoenolpyruvate carboxykinase (see Table 1). They presume that pyruvate formation from glucose is of minor importance, because whatever pyruvate is formed is probably reduced to lactate. From the labelling experiments it is clear that succinate predominates over lactate. The assumption, however, that low lactate production would mean low pyruvate production and therefore that carboxylation of pyruvate would play only a minor role, if any, is not correct. This conclusion is more or less in contradiction to the results of a later study in which the distribution of radioactive label was followed from L-[U-^{14}C]glucose incubated with mantle of *R. cuneata* (Stokes & Awapara, 1968). Here the authors concluded that about half of the consumed glucose could be accounted for by alanine, which was believed to be formed from pyruvate by a transamination reaction.

In contrast with Simpson & Awapara (1964), Jodrey & Wilbur (1955) obtained positive results for pyruvate carboxylase in oyster mantle. The presence of phosphoenolpyruvate carboxykinase, pyruvate carboxylase and 'malic' enzyme has also been demonstrated in mantle, hepatopancreas and adductor muscle of the sea mussel *M. edulis*. As reported for other bivalves, the specific activity of phosphoenolpyruvate carboxykinase was considerably higher than in rat liver. 'Malic' enzyme activity was extremely low (de Zwaan & van Marrewijk, 1973*a*; van Marrewijk *et al.*, 1973). Phosphoenolpyruvate carboxykinase and 'malic' enzyme were also detected in muscle and mantle tissue of *Placopecten magellanicus*, but no pyruvate carboxylase activity was found (O'Doherty & Feltham, 1971).

Kinetic studies on 'malic' enzyme from adductor muscle of *C. virginica* show that *in vivo* this enzyme functions mainly in the direction of pyruvate production. The optimum pH for the forward reaction (decarboxylation) is about 8.0 and for the reverse reaction about 5.3. The apparent K_m for pyruvate is about 15 mM and for malate 0.5 mM (Hochachka & Mustafa, 1973). Approximately the same results have been obtained for mantle and adductor muscle of *M. edulis* (R. A. F. M. Wijnands & A. de Zwaan, unpublished work). This fact also illustrates that the extremely low activity reported previously for *M. edulis* was due to the direction in which the test was performed (pH 7.6 and with 10 mM-pyruvate). The same is probably true for all other studies on bivalves in which no or low activities were obtained; in all these cases the enzyme assay was carried out in the direction of NADPH oxidation. Therefore it seems that CO_2 incorporation into pyruvate to form malate by the action of 'malic' enzyme as suggested by Hammen (1966) can be ruled out. The possibility of pyruvate carboxylation by pyruvate carboxylase has been generally ruled out because of the reported absence of this enzyme from bivalves (Chen & Awapara, 1969; O'Doherty & Feltham, 1971; Mustafa, 1976). On the other hand there are other reports which suggest the enzyme is present (Jodrey & Wilbur, 1955; de Zwaan *et al.*, 1973). The contradictory results obtained for oyster show that a negative test does not always mean the absence of the enzyme. The reported absence of pyruvate carboxylase from *R. cuneata* (Simpson & Awapara, 1964) seems to be in contrast with the radioactive tracer data obtained in a study on glucose formation from pyruvate using [2-^{14}C]pyruvate and [3-^{14}C]aspartate (Simpson & Awapara, 1965). The results indicated that indirect conversion of pyruvate into phosphoenolpyruvate via carboxylation of pyruvate and subsequent conversion of oxaloacetate into phosphoenolpyruvate, is the most favourable pathway. The involvement of 'malic' enzyme operating in the direction of CO_2 fixation seems unlikely and this experiment, therefore, provides indirect evidence for the presence of pyruvate carboxylase.

Table 1. Ratios of the activities of pyruvate kinase to the activities of phosphoenolpyruvate carboxykinase (a) and malate dehydrogenase to lactate dehydrogenase (b) and the relative importance of succinate and lactate as fermentation products

Species	Tissue	(a)	(b)	Succinate/lactate	Literature
Bivalves					
R. cuneata	Mantle	0.9	104	suc≫lac	Simpson & Awapara (1966)
C. virginica	Mantle	23	210	suc≫lac	Simpson & Awapara (1966)
Volsella demissus	Mantle	7	80	suc≫lac	Simpson & Awapara (1966)
P. magellanicus	Mantle	0.8	132		O'Doherty & Feltham (1971)
Mer. mercenaria	Mantle	Phosphoenolpyruvate carboxykinase negligible	—		Engel & Neat (1970)
M. edulis	Mantle	3.7	—		de Zwaan (1971)
P. magellanicus	Muscle	0.8	139		O'Doherty & Feltham (1971)
M. edulis	Muscle	1.3	71	suc≫lac (total animal)	de Zwaan (1971)
		1.3	—	suc≫lac*	van Marrewijk et al. (1973)
Cardium edule	Muscle	0.8	—		Gäde & Zebe (1973)
Mya arenaria	Muscle	13.5	—		Gäde & Zebe (1973)
An. cygnea	Muscle	1.3	—		Gäde & Zebe (1973)
Dreissena polymorpha	Muscle	0.6	—		Gäde & Zebe (1973)
Unio species	Muscle	1.5	—		Gäde & Zebe (1973)
Cardium edule	Foot	11.3	—		Gäde & Zebe (1973)
Mer. mercenaria	Foot	33.7	—		Engel & Neat (1970)
Mya arenaria	Siphon	3.1	—		Gäde & Zebe (1973)
M. edulis	Hepatopancreas	3.9	Lactate dehydrogenase negligible		de Zwaan (1971)

			Reference	End products
Parasitic helminths				
A. lumbricoides	0.04	37	Bueding & Saz (1968)	suc≫lac
Hymenolepis diminuta	0.18	6.6	Bueding & Saz (1968)	suc>lac
Schistosoma mansoni (o)	5.0	0.25	Bueding & Saz (1968)	lac≪suc
Trichinella spiralis (larvae)	0.3	—	Ward et al. (1969)	
Fasciola hepatica	—	—	Pritchard & Schofield (1968)	suc≫lac($-O_2$)
Moniezia expansa	Pyruvate kinase negligible	1.7	Bryant (1972)	suc≪lac($+O_2$)
Vertebrates				
Orconectes limosus	10.8	—	Gäde & Zebe (1973)	
Rana temporaria	6.1	—	Gäde & Zebe (1973)	
Pseudemys scripta elegans	9.3	—	Penny & Kornecki (1973)	
Liver	3.6	1.7	Simpson & Awapara (1966)	lac≫suc†
Heart	201		O'Doherty & Feltham (1971)	
Liver	466		Scrutton & Utter (1968)	
Chicken Muscle	7800	2.9	O'Doherty & Feltham (1971)	lac≫suc
Rat Muscle	29		de Zwaan (1971)	
Muscle	12		de Zwaan (1971)	
Liver	7		Scrutton & Utter (1968)	

* Gäde et al. (1975).
† Reeves (1963).

It seems therefore reasonable to conclude that the three CO_2-fixing enzymes phosphoenolpyruvate carboxykinase, pyruvate carboxylase and 'malic' enzyme have a general distribution among bivalves.

The 'malate' route introduced by Hammen (1966, 1969) and his scheme for glucose fermentation in the oyster is almost identical with a scheme presented by Saz & Weil (1962) for *A. lumbricoides*. Succinate is also a major fermentation product of this parasite. Saz & Weil (1962) also suggested that glucose is degraded to pyruvate, which is converted into malate by an NAD^+-dependent 'malic' enzyme. Recently this scheme underwent modification, because pyruvate kinase activity appeared to be almost absent (Saz & Lescure, 1969; see also Table 1). In the new scheme, phosphoenolpyruvate is carboxylated instead of pyruvate.

According to Bueding & Saz (1968) the metabolic fate of phosphoenolpyruvate is determined by the ratio of the activities of pyruvate kinase and phosphoenolpyruvate carboxykinase. This conclusion was based on a study of three parasitic helminths. There was a good correlation between the value of this ratio and the end products formed (derived either from oxaloacetate or pyruvate) (see Table 1 for *A. lumbricoides*, *Hymenolepis diminuta* and *Schistosoma mansoni*). For *Monienza expansa*, also a parasitic helminth, the ratio pyruvate kinase/phosphoenolpyruvate carboxykinase is 1.7, but this parasite produces lactate in air (phosphoenolpyruvate $\rightarrow$ pyruvate), but succinate under nitrogen (phosphoenolpyruvate $\rightarrow$ oxaloacetate). The switch from one end product to another is closely related to the properties of pyruvate kinase (Bryant, 1972). The ratio pyruvate kinase/phosphoenolpyruvate carboxykinase in bivalves is not indicative of the fate of phosphoenolpyruvate as shown by the data in Table 1. For example, in *M. edulis* and *An. cygnea* the ratio is nearly 1 and for *Cardium edule* nearly 10, but in all species succinate predominates over lactate as an anaerobic end product. The physiological meaning of the variations in the pyruvate kinase/phosphoenolpyruvate carboxykinase ratios (0.6–33.7 in Table 1) is not clear. Perhaps they are a reflexion of the importance of the succinate pathway during aerobic conditions. Table 1 shows that the extremes are 0.04 and 7800 for *A. lumbricoides* and rat muscle respectively. *Ascaris* has almost no pyruvate kinase activity and rat muscle almost no phosphoenolpyruvate carboxykinase activity. In this case the ratio is strongly indicative of the end products formed, but when the value ranges between 1 and 40, modulation of enzyme activities for one or both enzymes involved during anaerobiosis will strongly influence the metabolic fate of phosphoenolpyruvate.

The highest specific activities of phosphoenolpyruvate carboxykinase in bivalves are found in glycolytic tissue (i.e. adductor muscle; de Zwaan & de Bont, 1975), but in the rat the highest activities are found in gluconeogenic tissue (i.e. liver, kidney cortex; Scrutton & Utter, 1968). This is in agreement with a specific role for phosphoenolpyruvate carboxykinase in glycolytic fermentation of bivalves.

The situation in bivalves is not similar to *A. lumbricoides*, because of the high pyruvate kinase activities. It is therefore remarkable that pyruvate kinase activity is greatly decreased during anaerobiosis and this means that the ratios presented in Table 1 for bivalves gradually decrease (see under 'Phosphoenolpyruvate branchpoint'). The decreased rate of conversion of phosphoenol-pyruvate into pyruvate is a strong argument for Awapara & Campbell's (1964) supposition that CO_2 is incorporated into phosphoenolpyruvate. The involvement of pyruvate carboxylase in carboxylation instead of phosphoenolpyruvate carboxykinase also seems less favourable, because the former is entirely located in the mitochondrion (de Zwaan & van Marrewijk, 1973). For reasons of redox balance it is necessary for oxaloacetate to be available to the cytoplasmic malate dehydrogenase. This enzyme is thought to replace lactate dehydrogenase in reoxidizing cytoplasmic NADH and this may explain why the malate dehydrogenase/lactate dehydrogenase ratios are always high in species that form succinate (see Table 1).

From the above discussion it seems reasonable to suppose that phosphoenol-pyruvate rather than pyruvate is carboxylated in all species forming succinate; oxaloacetate is then reduced by malate dehydrogenase. This reaction is strongly exergonic ($\Delta G^{0'} = -6.7$ kcal; Lehninger, 1970), which favours the formation of malate. The dehydration of malate by fumarase is endergonic ($\Delta G^{0'} = +0.88$ kcal; Lehninger, 1970), but can proceed in this direction when fumarate is constantly removed. This occurs because the reduction of fumarate by NADH is highly exergonic ($\Delta G^{0'} = -14$ kcal) and therefore irreversible under normal physiological conditions (Chen & Awapara, 1969).

Schemes for glycolytic fermentation. In the route from glucose to succinate one oxidation (glyceraldehyde 3-phosphate $\rightarrow$ 1,3-diphosphoglycerate) and two reductions (oxaloacetate $\rightarrow$ malate; fumarate $\rightarrow$ succinate) are involved. Formation of NADH during glycolysis is balanced by the malate dehydrogenase reaction. The question remains, however, as to the origin of the NADH required for the fumarate reduction. Many attempts have been made to answer this question and various schemes have been proposed for the anaerobic degradation of carbohydrates and the maintenance of the redox balance.

Fig. 1 has been proposed by Awapara and co-workers (Stokes & Awapara, 1968; Chen & Awapara, 1969; Chen, 1969). For every molecule of glucose two molecules of phosphoenolpyruvate are formed. One molecule of phosphoenolpyruvate oxidizes two molecules of NADH on its way to succinate. For the production of every molecule of phosphoenolpyruvate only one molecule of

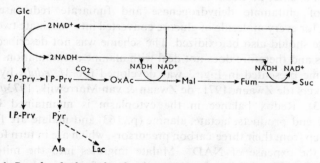

Fig. 1. *Postulated scheme for glycolysis and succinate production in R. cuneata*

The scheme is taken from Chen (1969). Abbreviations: *P*-Prv, phosphoenolpyruvate; Glc, glucose; OxAc, oxaloacetate; Mal, malate; Fum, fumarate; Suc, succinate; Lac, lactate.

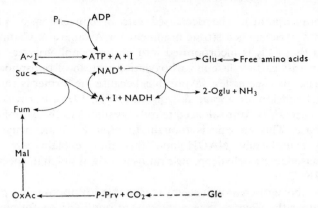

Fig. 2. *Possible mechanism for the maintenance of redox balance in succinate-accumulating invertebrates*

The scheme is taken from Gilles (1970). Oglu, oxoglutarate.

NAD^+ is reduced. Therefore these workers propose that the other molecule of phosphoenolpyruvate is used in some other conversion not involving an oxido-reduction step. This molecule should appear as pyruvate or some pyruvate derivative. From the experimental data they concluded that this derivative was alanine, which is formed from pyruvate in a transamination reaction with glutamate. Redox balance will be maintained only if succinate and alanine accumulate in equimolar amounts and when these compounds together account for almost all the degraded glucose. Fumarate is supposed to migrate into the mitochondrion and redox balance is maintained between the cytoplasm and mitochondrion, because the NADH for the fumarate reductase step is generated in the cytoplasm by the glyceraldehyde 3-phosphate dehydrogenase reaction. The authors suppose that mitochondria of *R. cuneata* are permeable to NADH (Chen & Awapara, 1969).

Gilles (1970) presented a scheme in which glucose degradation is related to amino acid oxidation via glutamate (Fig. 2). The NADH for the reduction of fumarate is generated by the oxidation of glutamate to 2-oxoglutarate and NH_3. Redox balance in the mitochondrion therefore depends on a simultaneous operation of glutamate dehydrogenase and fumarate reductase and this implies that for every molecule of glucose converted into succinate two molecules of glutamate should also be oxidized. The scheme was not described especially for molluscs and therefore does not deal with alanine accumulation.

The scheme presented in Fig. 3 was developed for *M. edulis* by de Zwaan and co-workers (de Zwaan, 1971; de Zwaan & van Marrewijk, 1973a; de Zwaan et al., 1973). Redox balance in the cytoplasm is maintained because all cytoplasmic end products, lactate, alanine (p. 163) and malate are formed in a reductive step from their three-carbon precursors, which are in turn formed from glucose at the expense of NAD^+. Malate migrates into the mitochondrion and is partly transformed into succinate via part of the tricarboxylic acid cycle operating in a reverse direction and partly into glutamate via 2-oxoglutarate in another part of the tricarboxylic acid cycle operating in the normal direction.

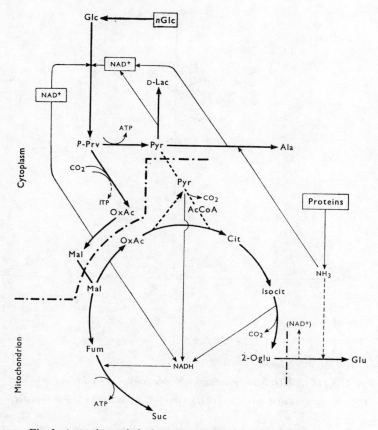

Fig. 3. *Anaerobic carbohydrate degradation in the sea mussel M. edulis*

The scheme is taken from de Zwaan (1971) and de Zwaan *et al.* (1973). Cit, citrate; Isocit, isocitrate.

This route involves one reduction (fumarate → succinate) and three oxidations (malate → oxaloacetate, pyruvate → acetyl-CoA and isocitrate → 2-oxoglutarate). It is implied that redox balance is obtained when the ratio of succinate and glutamate formation approximates to 3:1. The isotope distribution from [^{14}C]-glucose in succinate and glutamate was 19:7, which the authors considered to be evidence for this supposition (de Zwaan & van Marrewijk, 1973*a*). In this scheme redox balance is maintained separately in the cytoplasm and the mitochondrion. Succinate and alanine are supposed to accumulate in different compartments and their accumulation therefore occurs independently.

In 1972 Gilles presented a specific scheme for glucose degradation in molluscs (Fig. 4). All glucose is converted into malate and at this stage there is a redox balance. Then half of the malate is reduced to succinate and half is oxidized to alanine via pyruvate. The malate oxidation occurs by means of an NAD$^+$-dependent 'malic' enzyme, and the generated NADH is used to reduce fumarate to succinate. Redox balance depends on a simultaneous production of alanine and succinate, which together account for all the glucose. The conversion

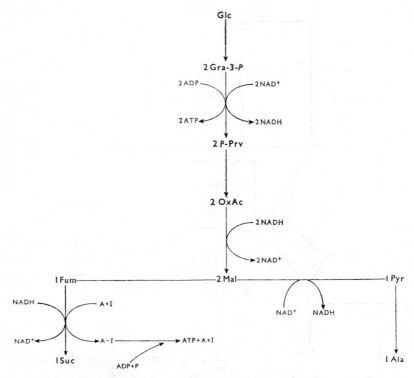

Fig. 4. *NAD⁺/NADH balance during anaerobic glucose degradation in molluscs*

The scheme is taken from Gilles (1972). Gra-3-*P*, glyceraldehyde 3-phosphate.

of pyruvate into alanine should not occur by a redox reaction. The difference between this scheme and that of Awapara and co-workers (Stokes & Awapara, 1968; Chen & Awapara, 1969; Chen, 1969) (Fig. 1) is that pyruvate kinase is not involved and redox balance is maintained separately on both sides of the mitochondrial membranes.

A scheme which can be regarded as a synthesis of the proposal of Awapara and co-workers and Gilles (1970) was presented by Hochachka & Mustafa (1972) (Fig. 5). Again redox balance is based on the simultaneous degradation of glucose and glutamate, but this is also coupled to a simultaneous accumulation of succinate and alanine in a ratio 2:1. This coupling takes place because of the fact that the function of glutamate dehydrogenase in generating NADH is replaced by 2-oxoglutarate dehydrogenase. An initial transamination of glutamate with pyruvate (→alanine) supplies the enzyme with its substrate. The formation of pyruvate from glucose is either via the pyruvate kinase pathway (as in Awapara's and de Zwaan's proposal) or via 'malic' enzyme [as in Gilles' (1972) proposal]. In this scheme half the succinate arises from glucose and half from glutamate. For every molecule of glucose one molecule of glutamate is used as in the proposal of Gilles (1970).

A problem in this scheme is that the formation of pyruvate via the malic' enzyme route will finally bring all the NADP⁺ into a reduced form.

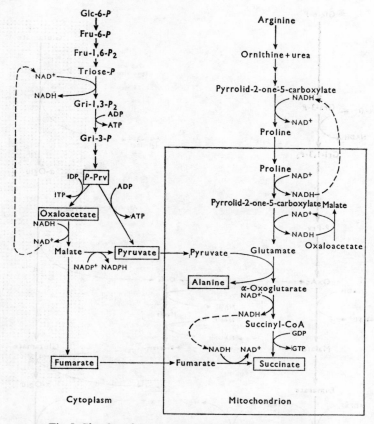

Fig. 5. *Glycolytic fermentation in facultative anaerobic animals*

The scheme is taken from Hochachka & Mustafa (1972). Abbreviations: Fru-6-P, fructose 6-phosphate; Fru-1,6-P_2, fructose 1,6-diphosphate; Triose-P, triose phosphate; Gri-1,3-P_2, 1,3-diphosphoglyceric acid; Gri-3-P, 3-phosphoglyceric acid.

Hochachka & Mustafa (1972) suggest that this might be overcome by reoxidation in reductive biosynthesis, such as fatty acid synthesis. In a later paper the scheme was reassessed (Hochachka *et al.*, 1973), as is shown in Fig. 6. Pyruvate is formed from aspartate via oxaloacetate and malate. The reoxidation of NADPH takes place by coupling the malate dehydrogenase and 'malic' enzyme reactions. This involves the action of a transhydrogenase. Redox balance depends now on a simultaneous mobilization of one molecule of glucose, two molecules of glutamate and two molecules of aspartate, and is again based on a 1:1 organization of 2-oxoglutarate dehydrogenase and fumarate reductase.

Fig. 7 shows a scheme for the sea mussel *M. edulis* not published before. This is an extension of the scheme presented previously in Fig. 3, but is based on new experimental data obtained by de Zwaan *et al.* (1975*a*), Addink & Veenhof (1975) and Kluytmans *et al.* (1975). It has been shown that the tricarboxylic acid cycle does not necessarily stop at the 2-oxoglutarate → glutamate stage, but can carry on to oxaloacetate and, therefore, it is suggested

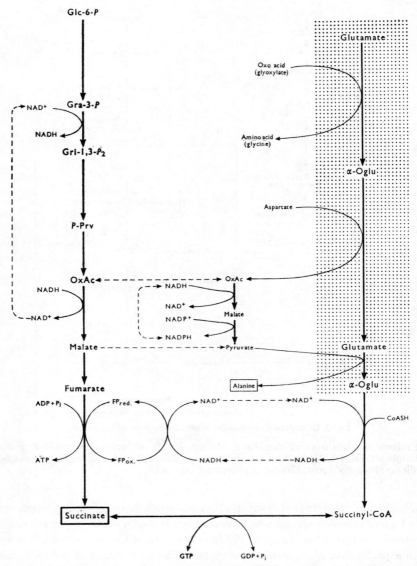

Fig. 6. *A metabolic map accounting for succinate and alanine accumulation as end products of anaerobic carbohydrate and amino acid catabolism*

The map is taken from Hochachka *et al.* (1973).

that the complete tricarboxylic acid cycle running in the forward direction generates the NADH for the reduction of fumarate to succinate (de Zwaan *et al.*, 1975a). With respect to the cytoplasmic reactions the scheme is identical with that in Fig. 3 and again malate (or fumarate) migrates into the mitochondrion. The cytoplasmic and mitochondrial parts of the overall glycolytic fermentation will probably occur in different organs (Addink & Veenhof, 1975) and this introduces a function for the circulatory system. In the mitochondrion malate

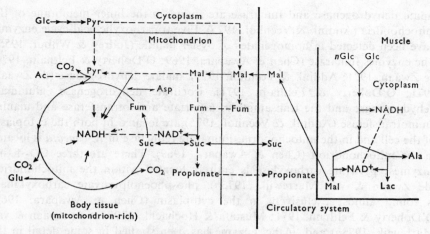

Fig. 7. *Anaerobic carbohydrate degradation in the sea mussel M. edulis*

See the text for an explanation.

will be converted into fumarate, which serves as electron acceptor in the electron-transfer chain. In order to supply the tricarboxylic acid cycle with acetyl-CoA, relatively small amounts of glucose might be converted into pyruvate (via phosphoenolpyruvate and/or malate). However, production of acetyl-CoA does not necessarily rely on a carbohydrate source [see review by de Zwaan & Wijsman (1975) under 'Sources of Energy Storage']. Besides the well-known end products succinate, alanine and D-lactate, propionate and acetate are also included (Kluytmans et al., 1975).

Evaluation of the schemes. (a) Presence and localization of the enzymes involved. All of the schemes presented include parts of the Embden–Meyerhof–Parnas pathway, the tricarboxylic acid cycle and probably part of the electron-transport chain (the reduction of fumarate). Many enzymes involved in these pathways have already been discussed under 'Fumarate formation'. All other enzymes involved have been detected in various molluscs except for a transhydrogenase which is supposed to play a role in the conversion of aspartate into alanine (Fig. 6; Hochachka et al., 1973) and an NAD^+-dependent 'malic' enzyme as supposed by Gilles (1972).

Enzymes of the Embden–Meyerhof–Parnas pathway have been detected in *Physa halei* (Beams, 1963), *Mytilus californianus* and *Haliotus rufescens* (Bennett & Nakada, 1968), *M. edulis* (de Zwaan, 1971; Addink & Veenhof, 1975) and *P. magellanicus* (O'Doherty & Feltham, 1971). The enzymes glyceraldehyde 3-phosphate dehydrogenase, lactate dehydrogenase, octopine dehydrogenase, pyruvate kinase, phosphoenolpyruvate carboxykinase, alanine aminotransferase and citrate synthase have been detected in six species of bivalves (Gäde & Zebe, 1973). Hammen (1968) measured transaminase activities (aspartate aminotransferase, alanine aminotransferase) in tissue homogenates of various marine bivalve molluscs. In *M. edulis* most tricarboxylic acid-cycle enzymes, two transaminases (aspartate aminotransferase and alanine aminotransferase) and glutamate dehydrogenase are located in the mitochondrial matrix and suc-

6

cinate dehydrogenase and fumarase are found in the inner membrane of the mitochondria (Addink & Veenhof, 1975). Five tricarboxylic acid-cycle enzymes have been detected in homogenates of oyster mantle (Jodrey & Wilbur, 1955). The enzymes fumarase (Chen & Awapara, 1969; O'Doherty & Feltham, 1971; de Zwaan, 1971; Addink & Veenhof, 1975), malate dehydrogenase (de Zwaan, 1971; O'Doherty & Feltham, 1971), isocitrate dehydrogenase, glutamate dehydrogenase and the transaminases aspartate aminotransferase and alanine aminotransferase (Addink & Veenhof, 1975) are located in both the cytoplasm of the cell and in the mitochondria. Fumarate reductase in *R. cuneata* is located in the mitochondrion (Chen & Awapara, 1969). There are three CO_2-fixing enzymes: pyruvate carboxylase is entirely located within the mitochondrion (de Zwaan & van Marrewijk, 1973*b*), phosphoenolpyruvate carboxykinase is found almost exclusively in the cytoplasm (Chen & Awapara, 1969; O'Doherty & Feltham, 1971; Mustafa & Hochachka, 1973*a*; de Zwaan & van Marrewijk, 1973*b*) and 'malic' enzyme has been studied in some detail in the cytoplasm of the adductor muscle of *C. gigas* (Hochachka & Mustafa, 1973) and *M. edulis*, and in mantle mitochondria of *M. edulis* (R. A. F. M. Wijnands & A. de Zwaan, unpublished work).

Evidence for the presence of an electron-transfer chain similar to that in mammalian cells has been presented by various authors (Kawai, 1959; Tappel, 1960; Ryan & King, 1962; Mattisson & Beechey, 1966; Gilles, 1972).

Further indirect evidence is available to prove that all pathways and related enzymes will function in molluscs (i.e. identification of intermediates and results with radioactively labelled substrates), but these references will not be treated here.

(b) Role of alanine. A common feature of most of the redox schemes is that the accumulation of alanine and succinate are closely connected to each other. Alanine accumulation is an ultimate condition for the generation of NADH, which serves to reduce fumarate to succinate. In the scheme for *R. cuneata* (Fig. 1), NADH is generated in the glyceraldehyde 3-phosphate dehydrogenase step, in the scheme drawn up by Gilles (1972) (Fig. 4) NADH production is coupled to the 'malic' enzyme reaction and in both schemes suggested by Hochachka and co-workers (Hochachka & Mustafa, 1972; Hochachka *et al.*, 1973) (Figs. 5 and 6) alanine accumulation serves to generate 2-oxoglutarate which is oxidized by an NAD$^+$-dependent 2-oxoglutarate dehydrogenase. This dependence involves an accumulation of succinate and alanine in a fixed ratio, which should be maintained during the total period of anaerobiosis. According to Stokes & Awapara (1968), Gilles (1972) and Hochachka and co-workers (Hochachka & Mustafa, 1972; Hochachka *et al.*, 1973) the ratio in which alanine and succinate accumulate should approach values of 1:1, 1:1 and 1:2 respectively.

Surprisingly, evidence for alanine accumulation during anaerobiosis in molluscs is very poor. Only a few reports show significant increases in the concentration of alanine during anaerobiosis (de Zwaan & Zandee, 1972*b*; Gäde *et al.*, 1975; de Zwaan *et al.*, 1976). In most studies, incorporation of radioactivity into alanine from glucose has been determined, without measuring the actual amount produced. In spite of these facts it has been generally accepted for a long time that in bivalves alanine is a main fermentation product and this is reflected in most of the proposed schemes. It is probably due to the great influence of the paper of Stokes & Awapara (1968). Their paper describes a study *in vitro* in which pieces of mantle tissue of *R. cuneata* were incubated for 1 h under anaerobic conditions with L-[U-^{14}C]glucose. The radioactivity in the amino acid and organic acid fractions together accounted for at least 95% of all the radioactive water-soluble substances in the extract. Alanine contained most of the radioactivity present in the amino acid fraction, whereas aspartic acid and glutamic acid contained only a small portion of the total. The radioactivity of the organic (anionic) fraction was almost exclusively found in succinic acid. The percentages of alanine and succinate together could account for over 80% of the label and the ratio of incorporation in these components was very close to one. From these data Stokes & Awapara (1968) drew the conclusion that succinate and alanine were formed in equimolar amounts and together could account for all the degraded carbohydrates. However, apart from the fact that it is hazardous to draw conclusions for the total animal when only the mantle is studied, it should be realized that this kind of experiment is not adequate to allow such interpretations [see also de Zwaan & Wijsman (1975) under 'Stoicheiometry of Glycolysis'].

More recent studies have revealed that volatile fatty acids, mainly propionate and acetate, play a dominant role as end products of glycolytic fermentations in bivalves (Kluytmans *et al.*, 1975; Gäde *et al.*, 1975). A less dominant role of volatile acids, mainly acetate, was already known for freshwater gastropods (Mehlmann & von Brand, 1951; von Brand *et al.*, 1955). Propionate is probably formed from glucose via succinate and can be considered (besides succinate itself) as a reflexion of the amount of NADH that becomes oxidized by the fumarate reductase enzyme complex.

When *M. edulis* was kept dry for 48h under N_2, a significant increase in succinate (2.70 μmol/g wet wt.) and alanine (3.32 μmol/g wet wt.) was observed (de Zwaan & Zandee, 1972b). This result seemed in reasonable agreement with the concept of Stokes & Awapara (1968). However, the volatile fatty acids were not determined, and alanine and succinate together could only account for about 50% of the degraded glycogen. In a later experiment, again performed on *M. edulis*, the concentrations of alanine were compared after 24 and 48h of anaerobiosis. These appeared not to be significantly different (6.60 and 7.58 μmol/100mg dry matter respectively). This was a first indication that alanine probably was an initial end product and that the ratio in which succinate and alanine accumulate therefore would be time-dependent (de Zwaan & van Marrewijk, 1973a). In a later study, in which accumulation of volatile fatty acids was determined after a 72h period of anaerobiosis, the ratio of succinate to alanine accumulation was 1.5. Propionate was accumulated in amounts equal to succinate, and the ratio of alanine/succinate plus propionate was 2.8 (Kluytmans *et al.*, 1975).

A paper on *An. cygnea* strongly supports the concept of alanine (and lactate) as initial end products of anaerobiosis (Gäde *et al.*, 1975). Indirect support is also obtained from studies on the kinetic properties of pyruvate kinase and phosphoenolpyruvate carboxykinase (see under 'Phosphoenolpyruvate branch-

Table 2. *Concentrations of various carboxylic acids and amino acids in M. edulis after different periods of anaerobiosis (J. H. F. M. Kluytmans, A. M. T. de Bont & T. C. M. Wijsman, unpublished work)*

	Concentration (μmol/g dry wt.)			
Length of exposure (days) ...	0	2	4	6
Succinate	4.51	28.99	41.76	51.85
Propionate	0.81	19.92	40.62	39.09
Alanine	49	65	56	62
Glutamate	35	52	58	63
Aspartate	28	30	27	33
Arginine	44	48	43	—
Proline	26	33	23	—
Glycine	210	248	202	222
NH$_3$	52	52	121	—

point'). Alanine and lactate accumulation will decrease pyruvate kinase activity by product inhibition (alanine) and a fall in pH (lactate), in favour of the phosphoenolpyruvate–oxaloacetate route. Once succinate is formed, this will cause a further fall in pH (see also under 'Phosphoenolpyruvate branchpoint'). Another observation which indicates that succinate accumulation is independent of alanine production is that in a few experiments with *M. edulis* there was almost no accumulation of alanine at all (see for instance Table 2). In a study (de Zwaan *et al.*, 1975a) the concentrations of alanine of the control group and a group kept for 17h under experimental anaerobiosis were 4.88 and 5.56mM respectively.

Summarizing, there is every reason to suppose that alanine is not an (permanent) end product in the sense of succinate and propionate. Accumulation of succinate and propionate are reflexions of an essential function, namely the operation of a modified electron-transfer chain. Energetically it is desirable to channel all glucose towards this system, but this needs an initial period to create conditions *in vivo* which stop the competing (aerobic) route for phosphoenol-pyruvate, and accumulation of alanine is one of the factors involved in the change.

(c) *Role of glutamate.* Another common feature of most schemes is that the formation of alanine is linked to the transamination of pyruvate and glutamate. As a result the amount of glutamate consumed will be equal to the accumulation of succinate (Stokes & Awapara, 1968; Gilles, 1972) or equal to half the accumulation of succinate (Hochachka & Mustafa, 1972; Hochachka *et al.*, 1973). When propionate is derived from succinate, an equivalent amount of glutamate will also be consumed. According to the scheme suggested by Hochachka & Mustafa (1972) and Hochachka *et al.* (1973) the same will be true for aspartate (Fig. 6). It is proposed that glutamate is obtained from the free amino acid pool (Chen & Awapara, 1969; Hochachka *et al.*, 1973). Steady-state concentrations of free amino acids in marine bivalves are indeed high (Allen, 1961; Florkin, 1966; Lynch & Wood, 1966; Schoffeniels & Gilles, 1972; Florkin & Bricteux-Grégoire, 1972; de Zwaan & van Marrewijk, 1973a). In *M. edulis* concentrations of glutamate and aspartate are about 35 and 30 μmol/g dry weight respectively (Table 2). These amounts would be sufficient to supply amino acids for anaerobiosis according to the schemes in Figs. 1, 4, 5 and 6 for 1–3 days. However, all data available at present show that during this period concentrations of glutamate and aspartate do not change significantly (de Zwaan & van Marrewijk, 1973a; de Zwaan et al., 1975a) not even when anaerobic conditions last for another few days and alanine accumulation reaches a standstill (Table 2). This shows that: (1) the fall in alanine production is not due to depletion of the glutamate stores and (2) the accumulation of succinate (and propionate) is unlikely to rely on glutamate consumption. As an objection to the second conclusion it could be assumed that besides degradation there will be a continuous formation of aspartate and glutamate. This has been proposed for glutamate by Hochachka & Mustafa (1972) and in their original scheme arginine and proline were considered especially as potential sources for glutamate formation (see Fig. 5). In the second scheme (Fig. 6) the position of proline as a substrate for anaerobiosis has been changed into an end product (Hochachka et al., 1973). As far as M. edulis is concerned neither proline nor arginine concentrations undergo substantial changes during anaerobiosis (Table 2) and these data therefore do not support their suggestion that the formation of proline as an end product of anaerobiosis would occur via glutamate. Although injection of [^{14}C]carbonate (Loxton & Chaplin, 1973), [^{14}C]glucose (Stokes & Awapara, 1968; de Zwaan & van Marrewijk, 1973a; de Zwaan et al., 1975a) and [^{14}C]glutamate (de Zwaan et al., 1975a) in bivalves always shows incorporation of radioactive label into glutamate, this is never true for proline.

Apart from the suggestions by Hochachka & Mustafa (1972) and Hochachka et al. (1973) no schemes in which glutamate donates its amino group to pyruvate deals with the fate of the 2-oxoglutarate. As discussed above this substrate is considered by Hochachka & Mustafa (1972) to be the hydrogen donor for the fumarate reduction.

Hochachka et al. (1973) point out that besides pyruvate glutamate could donate its amino group via a variety of transminase reactions:

$$\text{Oxo acid} + \text{glutamate} \rightarrow \text{2-oxoglutarate} + \text{amino acid}$$

The principal among these could be glycine aminotransferase. This was suggested because Cook et al. (1972) found high glycine concentrations in certain periods of the seasonal cycle in Balanus balanoides. Glycine accumulation certainly does not occur in M. edulis during anaerobiosis (de Zwaan & van Marrewijk, 1973a; Table 2). Moreover, this solution for the production of 2-oxoglutarate creates a new problem, namely what will be the source of the various oxo acids (i.e. glyoxylate for glycine accumulation).

It has also been suggested by Hochachka et al. (1973) that glutamate possibly derives ultimately from protein. This would lead to an increase in the overall concentration of free amino acids (the use of glutamate is compensated for by the production of alanine) or at least to a shift in the composition of the free amino acid pool. Neither occurs in M. edulis, except for some alanine accumulation (de Zwaan & van Marrewijk, 1973a; J. H. F. M. Kluytmans, A. M. T. de Bont & T. C. M. Wijsman, unpublished work). Studies on the anaerobic end products in freshwater snails (von Brand et al., 1950; Mehlmann & von Brand, 1951; de Zwaan et al., 1976), freshwater bivalves (Badman &

Chin, 1973; Gäde *et al.*, 1975) and marine bivalves revealed that there are only minor differences between the groups and we may therefore expect a similar basic mechanism for the carbohydrate fermentation. The fact that only marine bivalves maintain high concentrations of free amino acids also does not support a role for these compounds as an energy reserve during anaerobiosis.

Experiments on L-[U-^{14}C]glutamate degradation during anaerobiosis and aerobiosis showed that (1) the metabolism of glutamate was greatly decreased by anaerobiosis and (2) when glutamate was converted into 2-oxoglutarate and entered the tricarboxylic acid cycle it was not preferentially converted into succinate, but followed the cycle up to the stage of oxaloacetate (de Zwaan *et al.*, 1975a). The fact that in the metabolism of glutamate not only the oxidative step 2-oxoglutarate→succinyl-CoA is involved, but also the oxidative steps succinate→fumarate and malate→oxaloacetate, implies that the redox balance cannot be maintained according to the schemes of Figs. 5 and 6.

A similar study, performed on *Arenicola marina* revealed that, during anaerobiosis, hardly any metabolic conversion of [^{14}C]glutamate occurred (3% compared with 26% under aerobic conditions). Of the 3% converted into tricarboxylic acid-cycle intermediates, 41% appeared in succinate and 54% in malate (Zebe, 1975). A comparison with this non-parasitic annelid seems relevant here, because anaerobic glucose degradation showed great similarities to bivalves.

Summarizing, none of the experimental data support the idea that amino acids, especially glutamate, have a role as energy substrates during anaerobiosis.

(d) Role of the tricarboxylic acid cycle. Evidence for the operation of a normal tricarboxylic acid cycle under anaerobic conditions is based on experiments with radioactively labelled carbon substrates. Radioactivity from both glucose and glutamate was incorporated into 2-oxoglutarate, citrate, malate, fumarate, succinate and aspartate (and for glucose into glutamate) (de Zwaan & van Marrewijk, 1973a; de Zwaan *et al.*, 1975a). Radioactive CO_2 was also formed under anaerobic conditions. For glucose this could be produced by the pentose phosphate shunt, but with glutamate as the precursor this is sufficient evidence for the operation of, at least, part of the tricarboxylic acid cycle. The incorporation of radioactivity into CO_2 is greatly diminished during anaerobiosis (% ^{14}C in CO_2 falls from 22.9 to 2.0 and from 35.2 to 6.1 for glucose and glutamate respectively), and this shows that the flow of carbon through the tricarboxylic acid cycle is decreased. This is not surprising because many oxidation steps serve to produce NADH for the reduction of fumarate to succinate, and the rate of fumarate production will therefore be about five times that of the rate of the tricarboxylic acid cycle.

Anaerobic glutamate conversion was studied in *M. edulis* in which succinate had already accumulated in concentrations about twice that of aspartate before the injection of labelled glutamate (de Zwaan *et al.*, 1975a). A relatively high incorporation of labelled carbon was found in succinate and aspartate and the specific radioactivity in aspartate appeared to be twice that of succinate. Glutamate can only be converted into aspartate by the tricarboxylic acid cycle via succinate and oxaloacetate. The lower specific radioactivity in succinate can be understood by assuming that there are two metabolic pools of succinate. One pool would be formed by the conversion of glucose into succinate and although this is initially produced in the mitochondrion, it will diffuse into the cytoplasm and be distributed over all the body tissues. The second pool might be the steady-state concentration of succinate in its role as intermediate of the tricarboxylic acid cycle. A turnover between the two pools may explain why increasing amounts of carbon from glutamate will accumulate as succinate. The smaller succinate pool will have a much higher specific radioactivity than the total succinate pool and will be mainly involved in the production of aspartate from glutamate. This might explain the relatively high specific radioactivity of aspartate compared with the total succinate pool.

(e) Compartmentalization. One difficulty in the scheme shown in Fig. 7 is that the part of the tricarboxylic acid cycle between succinate and oxaloacetate operates in both directions. This problem can be solved if the forward and reverse sequences are spatially separated. Therefore it might be of significance that in bivalves malate dehydrogenase and fumarase are both located in the cytoplasm of the cell and in the mitochondrion (see above). Both the oxidation of succinate and the reduction of fumarate, are believed to occur in the mitochondrion, although not necessarily catalysed by the same enzyme (succinate dehydrogenase). It may be that succinate oxidation and fumarate reduction are catalysed by two separate enzymes, which have a different location within the mitochondrion. This also implies that fumarate instead of malate will pass the mitochondrial membranes, but this would be in contradiction with the general opinion that fumarate is a non-penetrant (Chappell, 1968). However, in the scheme for *R. cuneata* (Chen & Awapara, 1969) and in the scheme shown in Fig. 5 (Hochachka & Mustafa, 1972) it is supposed that fumarate migrates from the cytoplasm into the mitochondrion. Chen & Awapara (1969) indeed found that 'intact' mitochondria, when incubated with fumarate, accumulate succinate only when NADH is added to the incubation medium. This would suggest that fumarate and NADH can penetrate the mitochondrial membranes. Unfortunately, no convincing evidence was supplied to prove the intactness of the mitochondria, and moreover, there is some doubt about the correctness of the general concept that the mitochondrial membrane is impermeable to fumarate (Aleksandrowicz & Swierczynski, 1971).

The fact that Chen & Awapara (1969) consider the mitochondria to be permeable to NADH is their reason for supposing that the redox balance is maintained between the cytoplasm and the mitochondria. In the schemes of Hochachka & Mustafa (1972) and Hochachka *et al.* (1973) (Figs. 5 and 6) and de Zwaan (1971), de Zwaan & Marrewijk (1973a), de Zwaan *et al.* (1973) and de Zwaan *et al.* (1975a) (Figs. 3 and 7) redox balance is supposed to be maintained independently on both sides of the membrane barrier. An explanation for the differing conclusions about alanine formation (in the mitochondrion according to the first group and in the cytoplasm according to the second) is given under 'Phosphoenolpyruvate branchpoint'.

It is possible that the reaction sequence is not only divided between two compartments of the cell, but also amongst cells of different tissues. As shown by Addink & Veenhof (1975) the adductor muscle of *M. edulis* shows a high specific activity for the cytoplasmic enzymes, whereas no pyruvate dehydrogenase and 2-oxoglutarate dehydrogenase could be detected. This might be due to the low mitochondrial content of muscle tissue, in contrast with mantle and hepatopancreas. Addink & Veenhof (1975) therefore concluded that energy production in muscle is glycolytic, whereas the end products are oxidized in the hepatopancreas after transport through the haemolymph. The same can be supposed during anaerobiosis. For the muscle tissue there will be no serious difference, but in the absence of oxygen instead of the mitochondrial end products CO_2 and water, other products namely succinate, propionate and some CO_2 will be produced (Fig. 7). That malate and not lactate (as in the Cori cycle of mammals) will be the main intermediate circulating is in agreement with the proposition of van Weel (1974) that marine bivalves have lost the hepatic function and therefore the ability to reoxidize lactate from the muscles.

ATP Yield

The following reactions are possible sites of ATP production (see also the ATP Generation section).

(1) Substrate-level phosphorylations:

(a) 1,3-Diphosphoglycerate	→ 3-phosphoglycerate	(1 ATP)
(b) Phosphoenolpyruvate	→ pyruvate	(1 ATP)
(c) Phosphoenolpyruvate	→ oxaloacetate	(1 ATP)
(d) Succinyl-CoA	→ succinate	(1 ATP)
(e) Acetyl-CoA	→ acetate	(1 ATP)
(f) Methylmalonyl-CoA	→ propionyl-CoA	(1 ATP)
(g) Propionyl-CoA	→ propionate	(1 ATP)

(2) Electron-transfer-coupled phosphorylations:

(a) Fumarate	→ succinate	(1 ATP)
(b) NADH+reduced lipochrome	→ NAD^++oxidized lipochrome	(2 ATP)

For the reactions of (1e) and (1g) no evidence can be presented that these steps in molluscs are coupled with ATP synthesis, but theoretically it is possible that hydrolysis of these thiol esters is coupled with ATP synthesis (see Hochachka et al., 1973).

ATP synthesis coupled with the hydrolysis of acetyl-CoA has been shown for micro-organisms and is realized by two steps (Stadtman, 1966).

$$\text{Acetyl-CoA} + P_i \xrightarrow{(1)} \text{acetyl-}P + \text{CoA}$$
$$\text{Acetyl-}P + \text{ADP} \xrightarrow{(2)} \text{acetate} + \text{ATP}$$

$$\overline{\text{Acetyl-CoA} + P_i + \text{ADP} \longrightarrow \text{acetate} + \text{CoA} + \text{ATP}}$$

These reactions are catalysed by phosphotransacetylase (1) and acetate kinase (2).

In the following calculations of the number of ATP molecules produced per molecule of glucose it is supposed that all above-mentioned reactions are operative and that glucose units (as glucose 6-phosphate) are from glycogen degradation.

In the scheme of Stokes & Awapara (1968) 2 mol of ATP are generated in the phosphoglycerate kinase reaction, 1 in the phosphoenolpyruvate carboxykinase reaction and 1 in the fumarate-reduction step. Together 5 mol of ATP are generated and 1 mol of ATP is used (in the step catalysed by phosphofructokinase).

In the scheme presented by Hochachka et al. (1973) 2 mol of ATP are generated/mol of glucose in the phosphoglycerate kinase step, 2 mol/mol in the phosphoenolpyruvate carboxykinase step and two in the fumarate-reductase step. Another 2 mol of ATP are formed/2 mol of glutamate and 2 mol of aspartate (in the succinyl thiokinase reaction). These last 2 mol are very expensive from the point of needed substrate, because the same ATP production could be obtained if only another 0.4 mol of glucose (from glycogen) is converted into succinate. For bivalves, which are known to possess large glycogen reserves (Galtsoff, 1964; Giese, 1969; Walne, 1970; de Zwaan & Zandee, 1972b; Dare, 1973; Zs.-Nagy, 1973; Holland & Hannant, 1974; Gäde, 1975), the simultaneous mobilization of glucose and amino acids to gain extra ATP therefore seems illogical. Again 1 mol of ATP is used.

In the scheme of de Zwaan et al. (1975a) (Fig. 7) for the malate produced from glucose (the main route after an initial period of anaerobiosis) 4 mol of ATP are generated. Whatever the fate of malate (conversion into acetate, oxidation in the tricarboxylic acid cycle via pyruvate or reduction to succinate) 1 mol of ATP/mol of malate is always generated (or 2 from glucose). The main route of malate metabolism is clearly the conversion into succinate, and because succinate and propionate accumulate in the same proportion (J. H. F. M. Kluytmans, A. M. T. de Bont & T. C. M. Wijsman, unpublished work; Table 2) about another 2 mol of ATP will be generated per mol of glucose. Together almost 8 mol of ATP are generated per mol of glucose, 3 for succinate and 5 for propionate; 2 mol of ATP are used (in the steps fructose 6-phosphate→fructose 1,6-diphosphate and succinate→succinyl-CoA).

Summarizing, it is clear that all suggested modifications lead to a net increase in the ATP yield compared with the Embden–Meyerhof–Parnas pathway.

Advantages of the Modifications over the Embden–Meyerhof–Parnas Pathway

The following points may be considered as relevant physiological advantages. (1) There is an increased production of ATP. (2) The fact that the tricarboxylic acid cycle still operates allows the continuation of its role in anabolism and that other substances, besides carbohydrates, which can be converted into acetyl-CoA, may serve as energy reserves during anaerobiosis. The first fact may be of special importance to molluscs which undergo long-term anaerobiosis, e.g. Pisidium idahoense can survive 90 days of strictly anaerobic conditions (Cole, 1921). (3) Alanine formation may play a role in detoxification of NH_3 (see under 'Formation of alanine'). (4) The accumulation of alanine, succinate and propionate causes a smaller fall in pH than an equimolar amount of lactate. (5) Succinate, being an intermediate of the tricarboxylic acid cycle, can be readily oxidized on return of aerobic conditions and therefore the pH will return to normal again (and simultaneously the pyruvate kinase reduction). (6) Volatile acids may be more easily excreted than lactate in molluscs, which remain closed during anaerobiosis to avoid the danger of desiccation. This may contribute to maintaining the pH within a tolerable range for longer periods.

Regulation of Glycolytic Fermentation

Phosphoenolpyruvate branchpoint

In bivalve tissues that have significant quantities of both pyruvate kinase and phosphoenolpyruvate carboxykinase (Table 1) and in which the net flux of phosphoenolpyruvate depends on the metabolic conditions, the phosphoenolpyruvate branchpoint requires regulation. These conditions are gluconeogenesis (phosphoenolpyruvate → glucose), glycolytic fermentation (phosphoenolpyruvate → oxaloacetate) or aerobic oxidation (phosphoenolpyruvate → pyruvate). In the first two reaction sequences phosphoenolpyruvate has to be protected against conversion into pyruvate by pyruvate kinase. In both sequences phosphoenolpyruvate carboxykinase must be active, but in opposite directions. During aerobiosis, however, pyruvate kinase will predominate over phosphoenolpyruvate carboxykinase. Recently, many factors that can affect pyruvate kinase and phosphoenolpyruvate carboxykinase activity have been described.

Pyruvate kinase. The kinetics of pyruvate kinase have been studied on partially purified extracts of mantle and adductor muscle of *C. gigas* (Mustafa & Hochachka, 1971), adductor muscle of *M. edulis* (de Zwaan, 1971, 1972; de Zwaan & Holwerda, 1973; Holwerda & de Zwaan, 1973; Holwerda *et al.*, 1973; de Zwaan *et al.*, 1975b) and mantle of *M. edulis* (Livingstone & Bayne, 1974; Livingstone, 1975). In our laboratory we have also looked at pyruvate kinase of the adductor muscle of *Ostrea edulis*, *Mer. mercenaria*, *Chlamys operculatis*, *Pecten maximus*, the foot of *Patella vulgata* and the mantle, hepatopancreas and gill of *M. edulis*.

From the similarity in the behaviour of the enzymes studied towards the various effectors it seems that all studied enzymes probably possess co-operative characteristics. For instance, all enzymes show activation by fructose 1,6-diphosphate and this property is related to pyruvate kinases which exhibit classic allosteric behaviour of the type described by Monod *et al.* (1965) (see, for example, Hess *et al.*, 1966; Bailey & Walker, 1969; Llorente *et al.*, 1970; Jimenez de Asna *et al.*, 1971; Staal *et al.*, 1971). Therefore it seems that all tissues and species examined contain at least allosteric pyruvate kinases.

Co-operative behaviour (characterized by a simultaneous change in apparent K_m and the Hill coefficient) is observed in the binding of the substrate phosphoenolpyruvate and the effectors fructose 1,6-diphosphate, glucose 1,6-diphosphate, Mn^{2+}, alanine and ATP, but not for the co-substrate ADP. Fructose 1,6-diphosphate, glucose 1,6-diphosphate and Mn^{2+} are positive effectors, whereas alanine and ATP are negative effectors. Other positive effectors are cyclic AMP and OH^-, and negative effectors are high concentrations of ADP, phenylalanine, Ca^{2+} and H^+.

Of the positive effectors pyruvate kinase is by far the most sensitive to fructose 1,6-diphosphate. Its action is characterized by a shift of the phosphoenolpyruvate-saturation curve to a hyperbolic form with a substantial decrease in the concentration required for half-maximal velocity. Small amounts (μM) of fructose 1,6-diphosphate are able to reverse the effect of relatively high amounts (mM) of alanine and ATP.

Variation of pH has a pronounced effect on the activation by fructose 1,6-diphosphate and the inhibition by alanine and ATP. At lower pH, one gets a higher maximal activation than at higher pH, but more fructose 1,6-diphosphate is required. Livingstone & Bayne (1974) suggested that for the mantle enzyme, fructose 1,6-diphosphate activation would be independent of the pH. This was concluded from the equal ratio of $S_{0.5 \text{ phosphoenolpyruvate}}$ in the presence or absence of this effector. However, they determined the ratio only at two pH values, which were quite near to each other (7.2 and 7.5), and by combining their Figs. 2 and 8 it can be seen that the percentage stimulation by fructose 1,6-diphosphate at 7.2 was 140 and at 7.5, 110. pH also has a marked effect on $S_{0.5 \text{ phosphoenolpyruvate}}$ values in the absence of modulators. At lower pH the highest values are obtained. The shape of the pH-optimum curve is strongly affected by modulators. In the presence of fructose 1,6-diphosphate the activity changes only marginally within the physiological range, but in its absence there is a rapid fall on the acidic side of the optimum. This fall is increased by inhibitors.

All the observations on the effects of pH indicate that its action is dual: (1) an effect on charges and conformation of the enzyme protein and (2) an effect on the equilibrium between enzyme molecules in the relaxed and the tight (constrained) state. In the latter case the protons act in the same way as an allosteric inhibitor. It is striking that the influence of pH on the enzyme kinetics is opposite to that observed with allosteric (L-type enzymes after Tanaka et al., 1967) pyruvate kinases of mammalian tissues (Rozengurt et al., 1969; Staal et al., 1971; Black & Henderson, 1972). Also remarkable is the fact that the L-type enzymes are restricted to tissues with high gluconeogenic capacities (liver, kidney cortex), but in molluscs they are also distributed in typical glycolytic tissue such as the adductor muscle. These two factors are strong indications that the properties of pyruvate kinases in glycolytic tissue of molluscs are responsible for the modified glycolytic pattern and that the intracellular pH may have a dominant role in the control of the enzyme activities. A fall in pH will have two effects: (1) a direct inhibition of the catalytic capacity and (2) an indirect effect, because the potency of all allosteric inhibitors becomes stronger and that of the activators weaker.

In gluconeogenesis the same two enzymes, pyruvate kinase and phosphoenolpyruvate carboxykinase compete for phosphoenolpyruvate (not formed from carbohydrate precursors), and there is a gradual change in pyruvate kinase activity which follows the seasonal glycogen-storage cycle. Livingstone (1975) observed seasonal changes in $S_{0.5 \text{ phosphoenolpyruvate}}$ values for M. edulis, which were correlated with the gametogenic cycle (and therefore also with the glycogen cycle; see Gabbott, 1975). In spring and summer when glycogen accumulates,

Table 3. *Comparison of enzyme kinetics for phosphoenolpyruvate carboxykinase from the posterior adductor muscle of M. edulis and C. gigas*

The results for *M. edulis* are taken from de Zwaan & de Bont (1975) and those for *C. gigas* from Mustafa & Hochachka (1973*a,b*).

Characteristic	M. edulis	C. gigas
Ion requirement	Zn^{2+}, Mn^{2+}, Mg^{2+}	Zn^{2+}
Effect of Ca^{2+}	Inhibitor	—
pH optimum	6.0 (Zn^{2+})	5.1 (Zn^{2+})
	6.6 (Mn^{2+})	6.0 (Mn^{2+})
	6.7 (Mg^{2+})	
K_m for phosphoenolpyruvate	0.20, 0.20, 0.20	0.50, 0.20
	(Zn^{2+}; pH 6.0, 6.6, 7.3)	(Zn^{2+}; pH 5.1, 6.0)
	2.66, 0.23 (Mn^{2+}; pH 6.6, 7.3)	
	1.48, 0.46 (Mg^{2+}; pH 6.6, 7.3)	
Effect of ITP	Competitive inhibitor	Competitive inhibitor
Positive effector	H^+	H^+, alanine
Alanine reversal of ITP inhibition	—	+

high values of $S_{0.5 \text{ phosphoenolpyruvate}}$ were found. Gabbott (1975) discusses the possibility that these changes are related to the production of different isoenzymes.

Phosphoenolpyruvate carboxykinase. Kinetic studies on phosphoenolpyruvate carboxykinase have been restricted to the adductor muscle of *C. gigas* (Mustafa & Hochachka, 1973*a,b*) and *M. edulis* (de Zwaan & de Bont, 1975). In both studies the enzyme appeared to be of the non-allosteric type. The enzymes of both species are inhibited by ITP in a competitive way and activity is decreased at low pH. In spite of these similarities a number of differences were noted. This might be due, in some part, to the different way in which changes in absorbance were calculated. Table 3 summarizes the main results obtained by the two groups of workers.

One notable difference was that with Mg^{2+} no activity was obtained for *C. gigas* in contrast with *M. edulis*. Moreover, the pH optimum with Zn^{2+} for *C. gigas* was low compared with that for *M. edulis*. This implies that the two groups of workers have different opinions about the pH range in which the enzyme will play a significant role.

pH. From the above discussion it is clear that changes in pH will have opposite effects on the activities of phosphoenolpyruvate carboxykinase and pyruvate kinase. Therefore it is important to know if the pH changes during anaerobiosis and in which physiological range this occurs. Recently data have become available about pH values in relation to anaerobiosis (Crenshaw & Neff, 1969; Crenshaw, 1972; Wijsman, 1975). In the extrapallial fluid of various species a pH between 7.3 and 7.5 was observed, which decreased to a minimum of about 6.7 during shell closure. A lower initial pH was noticed for the foot of the sea mussel, namely about 7.0, which decreased to a minimum of about 6.5. The fall in pH in both the extrapallial fluid and the foot during the first 5h of shell closure was about 0.3 unit (Wijsman, 1975).

These studies show that the physiological pH range will be between 7.5 and 6.5. From their kinetic studies Mustafa & Hochachka (1971, 1973*a,b*) concluded that pyruvate kinase and phosphoenolpyruvate carboxykinase could not function at significant rates simultaneously and therefore an 'on/off' switch situation was envisaged for the phosphoenolpyruvate branchpoint. This was based mainly on their observation that the pH–activity profiles of both enzymes were essentially non-overlapping. For instance, at pH 6.0 adductor-muscle pyruvate kinase would be essentially inactive, whereas at pH 7.0 the affinity of pyruvate kinase for phosphoenolpyruvate is relatively high and phosphoenolpyruvate carboxykinase activity only 5% of the rate at optimal activity. Mustafa & Hochachka (1971; 1973a,b) clearly assumed that during anaerobiosis a pH below 6.5 would be reached. But the observations cited above suggest that there will be a gradual decrease in pH, but not below 6.5. According to Mustafa & Hochachka (1973a, b) the phosphoenolpyruvate carboxykinase route would never become switched on (at pH 6.5 the phosphoenolpyruvate carboxykinase activity is only about 25% of the rate at optimal activity), which is in contrast with the observed accumulation of succinate.

On the other hand Livingstone & Bayne (1974) and de Zwaan & de Bont (1975) concluded that both enzymes had overlapping pH profiles within the physiological range and therefore could proceed (compete) together and that the predominance of one reaction over the other would be determined by the degree of tissue anoxia or the length of anaerobiosis. This difference in opinion is mainly based on the different results obtained for phosphoenolpyruvate carboxykinase. In contrast with de Zwaan & de Bont (1975), Mustafa & Hochachka (1973a) consider Zn^{2+} to be the only natural cofactor for C. gigas, but for M. edulis Mg^{2+} might also be an important cofactor because of its pH profile and the dependence of the K_m for phosphoenolpyruvate on the pH (Table 3).

Formation of alanine. As was shown by Wijsman (1975) the fall in the pH diminishes gradually with the length of anaerobiosis. This observation fits very well with the concept of a simultaneous operation of both enzymes and the position of alanine as a temporary end product.

Alanine accumulation may be the result of a difference between the rate at which pyruvate is produced and the rate at which pyruvate can be channelled in the citric acid cycle. When bivalves close their shells there is an increase in tissue hypoxia, which gradually fades into complete anoxia (see under 'Use of stored oxygen'). As a first effect of decreased oxygen supply the rate of the citric acid cycle is diminished, but the pyruvate kinase reaction is only decreased after the pH decreases. In skeletal muscle a similar situation leads to lactate production, because the lactate dehydrogenase reaction by-passes the decreased capacity of the electron-transfer chain in reoxidizing cytoplasmic NADH. In molluscs, lactate dehydrogenase activities are very low and this may be the reason that alanine and not lactate accumulates. This implies that alanine is not formed from pyruvate by a transamination with glutamate, but via a redox reaction. The possibility of this conversion has been discussed by de Zwaan (1971). The reaction may depend on the direct action of an alanine dehydrogenase (reaction 3) or on an initial transamination, followed by reduction of 2-oxoglutarate by glutamate dehydrogenase (reactions 1 and 2).

$$\text{Pyruvate} + \text{glutamate} \quad\quad\quad\quad\quad\quad\quad \leftrightarrow \text{alanine} + \text{2-oxoglutarate} \quad\quad (1)$$
$$\text{2-Oxoglutarate} + NH_3 + NADH + H^+ \leftrightarrow \text{glutamate} + NAD^+ + H_2O \quad\quad (2)$$

$$\text{Pyruvate} + NH_3 + NADH + H^+ \quad\quad\quad \leftrightarrow \text{alanine} + NAD^+ + H_2O \quad\quad (3)$$

Direct evidence that this conversion occurs is not available, but as discussed above there is also no evidence that there is a net consumption of glutamate. Moreover, both the enzymes alanine aminotransferase and glutamate dehydrogenase are located in the cytoplasm and in the mitochondrion. In mammalian tissues, transaminases and glutamate dehydrogenase can operate in an enzyme–enzyme complex. Under conditions where alanine does not react at all with glutamate dehydrogenase, the addition of alanine aminotransferase results in alanine dehydrogenase activity (Fahien & Smith, 1974). It has been observed that during anoxia various amounts of NH_3 are available in the tissues of the sea mussel (see Table 2). NH_3 is the main end product of protein catabolism and is excreted into the seawater under aerobic conditions. The formation of alanine may therefore also be of significance because it results in the detoxification of NH_3 by fixation onto pyruvate.

Accumulation of CO_2 (produced in the tricarboxylic acid cycle especially as long as there is oxygen) and some initial lactate production may cause a fall in the pH and consequently a shift in the competition for phosphoenolpyruvate between phosphoenolpyruvate carboxykinase and pyruvate kinase in favour of phosphoenolpyruvate carboxykinase. The initial production of alanine will have the same effect, but in marine bivalves this is of minor importance, because the steady-state concentrations are already so high that about maximal inhibition is obtained.

Once succinate is formed, its accumulation will contribute to a further fall in pH. The pyruvate kinase step will be much more strongly inhibited than the phosphoenolpyruvate carboxykinase step will be activated (de Zwaan & de Bont, 1975). This implies that the total flux of substrate through both pathways will diminish, which again explains why normally no Pasteur effect is noticed in bivalves (de Zwaan & Wijsman, 1975). According to the latest concept of de Zwaan and co-workers (see Fig. 7) the citric acid cycle will continue even after complete anaerobiosis (fumarate replaces oxygen) at a decreased rate so that when a certain level of inhibition of pyruvate kinase is obtained there may be an equilibrium between the pyruvate kinase and the pyruvate dehydrogenase reactions which makes any further accumulation of alanine superfluous. There will always be some pyruvate production by the action of pyruvate kinase during anaerobiosis which will stay in equilibrium with alanine but without a net conversion. This may explain why after an initial period of 15h anaerobiosis, injection of L-[U-^{14}C]glucose still led to a reasonable incorporation of the ^{14}C label into alanine, whereas the actual amount did not increase (de Zwaan et al., 1975a).

In contrast with the formation of alanine in the cytoplasm from glucose via the pyruvate kinase route Hochachka & Mustafa (1972) and Hochachka et al. (1973) expect the alanine production to occur in the mitochondrion (Fig. 5). Pyruvate is supposed to be formed from malate (formed from glucose or aspartate) via 'malic' enzyme. This supposition is in agreement with their view about the regulation of the phosphoenolpyruvate branchpoint. As discussed above, this group considers the simultaneous accumulation of succinate and alanine to be a basic condition for the maintenance of redox balance. Their 'on/off' switch concept for the phosphoenolpyruvate branchpoint implies that succinate production means switching off the pyruvate kinase reaction, and because alanine should accumulate along with succinate this reaction cannot be involved in alanine formation. Recent studies also suggest a role for the 'malic' enzyme step in alanine formation in *C. gigas*. Of the total L-[U-^{14}C]alanine utilized after 1 h of anoxia 17% was accounted for by alanine which certainly means that at least some oxaloacetate formed either from aspartate or glucose can be converted into alanine (Collicut, 1975). Taking into consideration that the pH of the incubation medium was 7.4 and lower pH values are more favourable to this enzyme (Hochachka & Mustafa, 1973) this percentage will probably be even higher at lower pH values. The important role of acetate as end product in *An. cygnea* (Gäde *et al.*, 1975) may also be related to an active role for 'malic' enzyme in the metabolism of oxaloacetate.

In *M. edulis*, metabolism of L-[U-^{14}C]glutamate under anaerobic conditions results in a high percentage of the radioactive label in aspartate. This implies that malate also had a high specific radioactivity, but in spite of this fact incorporation of label into alanine was only 20% of that into aspartate. The data do not support the suggestion of an active role for 'malic' enzyme in the formation of alanine in *M. edulis* (de Zwaan *et al.*, 1975a). Summarizing, we may say that the main route for the formation of alanine is still uncertain and it may well be that at this point clear species differences occur.

Energy charge

Atkinson (1968) introduced the concept of energy charge, which is half of the number of anhydride-bound phosphate groups per adenosine. He considers this parameter to be a major regulatory factor at every point where metabolites are partitioned between energy-yielding and energy-demanding processes. It can only be important when the values are not too low. Wijsman (1976) has calculated the energy charge for various tissues of *M. edulis* and the relation to anaerobiosis in the whole animal. The control value of the various tissues ranged between 0.69 (hepatopancreas) and 0.91 (adductor muscle) and was 0.85 for the whole body tissue. During 7 days of anaerobiosis the whole body values ranged between 0.85 and 0.62. The values for *M. edulis* correspond to those for mammalian tissues. During anaerobiosis this parameter will be involved in the regulation of energy metabolism in the sea mussel, because the values never become too low. We are now studying how the activities of various key enzymes depend on the energy charge of the body tissues.

We thank Dr. P. A. Gabbott and Dr. T. C. M. Wijsman for their critical comments on this manuscript.

References

Addink, A. D. F. & Veenhof, P. R. (1975) *Proc. Eur. Mar. Biol. Symp. 9th*, 109–119
Aleksandrowicz, Z. & Swierczynski, J. (1971) *FEBS Lett.* **15**, 269–272
Allen, K. (1961) *Am. Zool.* **1**, 253–261
Atkinson, D. E. (1968) *Biochemistry* **7**, 4030–4034
Awapara, J. & Campbell, J. W. (1964) *Comp. Biochem. Physiol.* **11**, 231–235

Badman, D. G. & Chin, S. L. (1973) *Comp. Biochem. Physiol. B* **44**, 27–32
Bailey, E. & Walker, P. (1969) *Biochem. J.* **111**, 359–364
Beams, C. G., Jr. (1963) *Comp. Biochem. Physiol.* **8**, 109–114
Bennett, R. & Nakada, H. I. (1968) *Comp. Biochem. Physiol.* **24**, 787–797
Black, J. A. & Henderson, M. H. (1972) *Biochim. Biophys. Acta* **284**, 115–127
Bryant, C. (1972) *Int. J. Parasitol.* **2**, 333–340
Bueding, E. & Saz, H. J. (1968) *Comp. Biochem. Physiol.* **24**, 511–518
Chappell, J. B. (1968) *Br. Med. Bull.* **24**, 150–157
Chen, C. H. (1969) Ph.D. Thesis, Rice University, Houston
Chen, C. H. & Awapara, J. (1969) *Comp. Biochem. Physiol.* **30**, 727–737
Cole, A. E. (1921) *J. Exp. Zool.* **33**, 293
Collicutt, J. (1975) M.Sc. Thesis, University of British Columbia
Cook, P. A., Gabbott, P. A. & Youngson, A. (1972) *Comp. Biochem. Physiol. B* **42**, 409–421
Crenshaw, M. A. (1972) *Biol. Bull. (Woods Hole Mass.)* **143**, 506–512
Crenshaw, M. A. & Neff, J. M. (1969) *Am. Zool.* **9**, 881–889
Dare, P. J. (1973) *Coop. Res. Rep. Int. Counc. Explor. Sea*, pp. 1–6
de Zoeten, L. W. & Tipker, J. (1969) *Hoppe-Seyler's Z. Physiol. Chem.* **350**, 691–695
de Zwaan, A. (1971) Ph.D. Thesis, State University of Utrecht
de Zwaan, A. (1972) *Comp. Biochem. Physiol. B***42**, 7–14
de Zwaan, A. & de Bont, A. M. Th. (1975) *J. Comp. Physiol.* **96**, 85–94
de Zwaan, A. & Holwerda, D. A. (1973) *Biochim. Biophys. Acta* **276**, 430–433
de Zwaan, A. & van Marrewijk, W. J. A. (1973*a*) *Comp. Biochem. Physiol. B* **44**, 429–439
de Zwaan, A. & van Marrewijk, W. J. A. (1973*b*) *Comp. Biochem. Physiol. B* **44**, 1057–1066
de Zwaan, A. & Wijsman, T. C. M. (1975) *Comp. Biochem. Physiol.* in the press
de Zwaan, A. & Zandee, D. I. (1972*a*) *Comp. Biochem. Physiol. A* **43**, 53–58
de Zwaan, A. & Zandee, D. I. (1972*b*) *Comp. Biochem. Physiol. B* **43**, 47–54
de Zwaan, A., van Marrewijk, W. J. A. & Holwerda, D. A. (1973) *Neth. J. Zool.* **23**, 225–228
de Zwaan, A., de Bont, A. M. Th. & Kluytmans, J. H. F. M. (1975*a*) *Proc. Eur. Mar. Biol. Symp. 9th*, 121–138
de Zwaan, A., Holwerda, D. A. & Addink, A. D. F. (1975*b*) *Comp. Biochem. Physiol. B* **52**, 469–472
de Zwaan, A., Mohamed, A. M. & Geraerts, W. P. M. (1976) *Neth. J. Zool.* in the press
Engel, H. R. & Neat, M. J. (1970) *Comp. Biochem. Physiol.* **37**, 397–403
Fahien, L. A. & Smith, S. E. (1974) *J. Biol. Chem.* **249**, 2696–2703
Florkin, M. (1966) in *Physiology of Mollusca* (Wilbur, K. M. & Yonge, C. M., eds.), vol. 2, pp. 309–343, Academic Press, New York and London
Florkin, M. & Bricteux-Grégoire, S. (1972) *Chem. Zool.* **7**, 301–342
Gabbott, P. A. (1975) *Proc. Eur. Mar. Biol. Symp. 9th* 191–211
Gäde, G. (1975) Ph.D. Thesis, Westfälische Wilhelms-Universität, Münster
Gäde, G. & Grieshaber, M. (1975) *J. Comp. Physiol.* **102**, 149–158
Gäde, G. & Zebe, E. (1973) *J. Comp. Physiol.* **85**, 291–301
Gäde, G., Wilps, H., Kluytmans, J. H. F. M. & de Zwaan, A. (1975) *J. Comp. Physiol.* **104**, 79–85
Galtsoff, P. S. (1964) *Bur. Sport Fish Wildl. (U.S.) Res. Rep.* **64**
Ghiretti, F. & Ghiretti-Magaldi, A. (1972) *Chem. Zool.* **7**, 201–214
Giese, A. C. (1969) *Oceanograf. Mar. Biol. Annu. Rev.* **7**, 175–229
Gilles, R. (1970) *Arch. Int. Physiol. Biochim.* **78**, 313–326
Gilles, R. (1972) *Chem. Zool.* **7**, 467–495
Haas, D. W. (1964) *Biochim. Biophys. Acta* **92**, 433–439
Hammen, C. S. (1964) *Proc. Symp. Exp. Mar. Ecol.* **2**, 48–50
Hammen, C. S. (1966) *Comp. Biochem. Physiol.* **17**, 289–296
Hammen, C. S. (1968) *Comp. Biochem. Physiol.* **26**, 697–705
Hammen, C. S. (1969) *Am. Zool.* **9**, 309–318
Hammen, C. S. (1975) *Comp. Biochem. Physiol. B* **50**, 407–412
Hammen, C. S. & Lum, S. C. (1962) *J. Biol. Chem.* **237**, 2419–2422
Hammen, C. S. & Lum, S. C. (1966) *Comp. Biochem. Physiol.* **19**, 775–781
Hammen, C. S. & Osborne, P. J. (1959) *Science* **130**, 1409–1410
Hammen, C. S. & Wilbur, K. M. (1959) *J. Biol. Chem.* **234**, 1268–1271
Hess, B., Haeckel, R. & Brand, K. (1966) *Biochem. Biophys. Res. Commun.* **24**, 824–831
Hochachka, P. W. & Mustafa, T. (1972) *Science* **178**, 1056–1060
Hochachka, P. W. & Mustafa, T. (1973) *Comp. Biochem. Physiol. B* **45**, 625–637
Hochachka, P. W., Fields, J. & Mustafa, T. (1973) *Am. Zool.* **13**, 543–555
Holland, D. I. & Hannant, P. J. (1974) *J. Mar. Biol. Assoc. U.K.* **54**, 1007–1016

Holwerda, D. A. & de Zwaan, A. (1973) *Biochim. Biophys. Acta* **309**, 296–306
Holwerda, D. A., de Zwaan, A. & van Marrewijk, W. J. A. (1973) *Neth. J. Zool.* **23**, 232–235
Jimenez de Asna, L., Rozengurt, E. & Carminatti, H. (1971) *FEBS Lett.* **14**, 22–24
Jodrey, L. H. & Wilbur, K. M. (1955) *Biol. Bull. (Woods Hole, Mass.)* **108**, 346–358
Karnaukhov, V. N. (1971) *Exp. Cell Res.* **64**, 301–306
Kawai, K. (1959) *Biol. Bull. (Woods Hole, Mass.)* **117**, 125–132
Kluytmans, J. H. F. M. & de Zwaan, A. (1976) *Biochem. Soc. Trans.* in the press
Kluytmans, J. H. F. M., Veenhof, P. R. & de Zwaan, A. (1975) *J. Comp. Physiol.* **104**, 71–78
Lábos, E., Zs.-Nagy, I. & Hiripi, L. (1966) *Ann. Biol. Tihany* **33**, 37–44
Lehninger, A. L. (1970) *Biochemistry*, pp. 298 and 487, Worth Publishers, New York
Livingstone, D. R. (1975) *Eur. Mar. Biol. Symp. 9th* 151–164
Livingstone, D. R. & Bayne, B. L. (1974) *Comp. Biochem. Physiol. B* **48**, 481–497
Llorente, P., Marco, R. & Sols, A. (1970) *Eur. J. Biochem.* **13**, 45–54
Loxton, J. & Chaplin, A. E. (1973) *Biochem. Soc. Trans.* **1**, 419–421
Lynch, M. P. & Wood, L. (1966) *Comp. Biochem. Physiol.* **19**, 783–790
Mattisson, A. G. M. & Beechey, R. B. (1966) *Exp. Cell Res.* **41**, 227–243
Mehlmann, B. & von Brand, T. (1951) *Biol. Bull. (Woods Hole, Mass.)* **100**, 199–205
Monod, J., Wijman, J. & Changeux, J-P. (1965) *J. Mol. Biol.* **12**, 88
Mustafa, T. (1976) *Comp. Biochem. Physiol.* in the press
Mustafa, T. & Hochachka, P. W. (1971) *J. Biol. Chem.* **246**, 3196–3203
Mustafa, T. & Hochachka, P. W. (1973a) *Comp. Biochem. Physiol. B* **45**, 639–655
Mustafa, T. & Hochachka, P. W. (1973b) *Comp. Biochem. Physiol. B* **45**, 657–667
O'Doherty, P. J. A. & Feltham, L. A. W. (1971) *Comp. Biochem. Physiol. B* **38**, 543–551
Penny, D. G. & Kornecki, E. H. (1973) *Comp. Biochem. Physiol. B* **46**, 405–415
Prichard, R. K. & Schofield, P. J. (1968) *Comp. Biochem. Physiol.* **24**, 697–710
Ryan, C. A. & King, T. E. (1962) *Biochim. Biophys. Acta* **62**, 269–278
Reeves, R. B. (1963) *Am. J. Physiol.* **205**, 23–29
Rosen, B. (1966) *Fish. Ind. Res.* **3**, 5–11
Rozengurt, E., Jimenez de Asna, L. & Carminatti, H. (1969) *J. Biol. Chem.* **244**, 3142–3147
Sanadi, D. R. & Fluharty, A. L. (1963) *Biochemistry* **2**, 523–528
Saz, H. J. & Lescure, O. L. (1969) *Comp. Biochem. Physiol.* **30**, 49–60
Saz, H. J. & Weil, A. (1962) *J. Biol. Chem.* **237**, 2053–2056
Schoffeniels, E. & Gilles, R. (1972) *Chem. Zool.* **7**, 393–418
Scrutton, C. M. & Utter, M. F. (1968) *Annu. Rev. Biochem.* **37**, 249–302
Seidman, I. & Entner, N. (1961) *J. Biol. Chem.* **236**, 915–919
Simpson, J. W. & Awapara, J. (1964) *Comp. Biochem. Physiol.* **12**, 457–464
Simpson, J. W. & Awapara, J. (1965) *Comp. Biochem. Physiol.* **15**, 1–6
Simpson, J. W. & Awapara, J. (1966) *Comp. Biochem. Physiol.* **18**, 537–548
Singer, T. P. (1971) in *Biochemical Evolution and the Origin of Life* (Schoffeniels, E., ed.), pp. 203–223, North-Holland, Amsterdam
Staal, G. E. J., Koster, J. F., Kamp, H., van Milligen-Boersma, L. & Veeger, C. (1971) *Biochim. Biophys. Acta* **227**, 86–96
Stadtman, E. R. (1966) in *Current Aspects of Biochemical Energetics* (Kaplan, N. O. & Kennedy, E. P., eds.), pp. 39–62, Academic Press, New York and London
Stadtman, T. C. & Elliot, P. (1956) *J. Am. Chem. Soc.* **78**, 2020–2021
Stadtman, T. C., Elliot, P. & Tiemann, L. (1958) *J. Biol. Chem.* **231**, 961–974
Stokes, T. M. & Awapara, J. (1968) *Comp. Biochem. Physiol.* **25**, 883–892
Tanaka, T., Harano, Y., Sue, F. & Morimura, H. (1967) *J. Biochem. (Tokyo)*, **62**, 71–91
Tappel, A. L. (1960) *J. Cell Comp. Physiol.* **55**, 111–126
van Weel, P. B. (1974) *Comp. Biochem. Physiol. A* **47**, 1–9
von Brand, T. (1946) *Biodynamica* **4**
von Brand, T., Baerstein, H. D. & Mehlmann, B. (1950) *Biol. Bull. (Woods Hole, Mass.)* **98**, 266–276
von Brand, T., McMahon, P. & Nolan, M. O. (1955) *Physiol. Zool.* **28**, 35–40
Walne, P. R. (1970) *Fish. Invest. London Ser. 2* **26**, 1–33
Ward, C. S., Castro, G. A. & Fairbairn, D. (1969) *J. Parasitol.* **55**, 67–71
Wegener, B. A., Barnitt, A. E. & Hammen, C. S. (1969) *Life Sci.* **8**, 335–343
Wijsman, T. C. M. (1975) *Proc. Eur. Mar. Biol. Symp. 9th* 139–149
Wijsman, T. C. M. (1976) *J. Comp. Physiol.* in the press
Wilson, M. A. & Cascarano, J. (1970) *Biochim. Biophys. Acta* **216**, 54–62
Zebe, E. (1975) *J. Comp. Physiol.* **101**, 133–147
Zs.-Nagy, I. (1969) *Acta Biol. Acad. Sci. Hung.* **20**, 451–463

Zs.-Nagy, I. (1971a) *Ann. Biol. Tihany* **38**, 117–129
Zs.-Nagy, I. (1971b) *Comp. Biochem. Physiol. A* **40**, 595–602
Zs.-Nagy, I. (1973) *Acta Biochim. Biophys. Acad. Sci. Hung.* **8**, 143–151
Zs.-Nagy, I. (1974) *Comp. Biochem. Physiol. A* **49**, 399–405
Zs.-Nagy, I. & Csukàs, C. (1969) *Ann. Biol. Tihany* **36**, 115–122
Zs.-Nagy, I. & Ermini, M. (1972) *Comp. Biochem. Physiol. B* **43**, 593–600

Biochem. Soc. Symp. (1976) **41**, 169–178
Printed in Great Britain

Metabolic Consequences of Submersion Asphyxia in Mammals and Birds

By ARNOLDUS SCHYTTE BLIX

*Institute of Medical Biology, Section of Physiology, University of Tromsø,
Tromsø, Norway*

Synopsis

On submersion, the cardiovascular system of naturally diving animals is virtually transformed into a heart–brain–lung preparation as the result of intense and highly selective neurogenic vasoconstriction. A supply of oxygen from the circulating blood cells to the heart and the brain is thereby secured. Peripheral tissues, e.g. muscles and kidneys, have to depend on local stores of oxygen, i.e. myoglobin, or, when these are exhausted, on anaerobic metabolism. In spite of such physiological adjustments, however, arterial pO_2 will inevitably decrease throughout the underwater episode. In prolonged dives values lower than 10mmHg can be observed. Moreover, the ischaemia in the peripheral tissues will, in some animals, e.g. arctic seals and whales, result in a profound decrease in the temperature of tissues such as the skin. The aim of the present report is to review some of the more important biochemical mechanisms that allow diving mammals and birds to cope with the above-mentioned combined asphyxic and temperature stresses for prolonged periods. Some of the osmotic problems arising from the rapid wash-out of acidic metabolic end products from the previously ischaemic tissues on emergence are also discussed.

Introduction

On submersion most vertebrates meet definite metabolic problems. In natural divers, these are dealt with in three ways. The first is to have large oxygen stores, and the second is, by physiological reflex adjustments, to economize these stores. The third is by biochemical adaptations to improve the ability for anaerobic energy metabolism. As the biochemical and the physiological adaptations are equally important in explaining the metabolic consequences of diving, both aspects will be considered in this review. It is, however, not my intention to give a comprehensive survey of the literature on physiological adaptations to a diving habit. Only those studies which I found important for the understanding of the basic principles of how diving animals cope with the different metabolic consequences of submersion asphyxia are included. Most of the examples are taken from studies of different species of seals and ducks.

Oxygen Stores and Oxygen Budget

The blood volume and erythrocyte number are much larger in diving animals than in their terrestrial relatives (cf. Lenfant *et al.*, 1970). This gives the diving animals a greater blood oxygen capacity (i.e. a greater blood oxygen store) than the non-divers, but the high blood cell content results in an increased blood viscosity, which is further enhanced by decreased temperatures (see below) (Guard & Murrish, 1973).

The lung oxygen-storage capacity is small in diving animals (cf. Lenfant *et al.*, 1970), and the oxygen in the lung is very soon depleted during a dive (Andersen, 1959). The lung contribution to the total oxygen store is therefore of minor importance. In fact, most animals expire at the commencement of the dive.

In contrast, muscle myoglobin concentration is very high, and Robinson (1939) estimated the oxygen captured as oxymyoglobin to be as much as 47% of the total oxygen store. However, because of the physiological adjustments mentioned below, this oxygen store is restricted to the muscles and is therefore unimportant in increasing diving time.

Quantitative comparisons between the amount of O_2 available from the above-mentioned depots and the resting oxygen consumption of the animal have been carried out for several species (see, e.g., Scholander, 1940). Such calculations have, without exception, shown that the diving vertebrates are actually able to stay

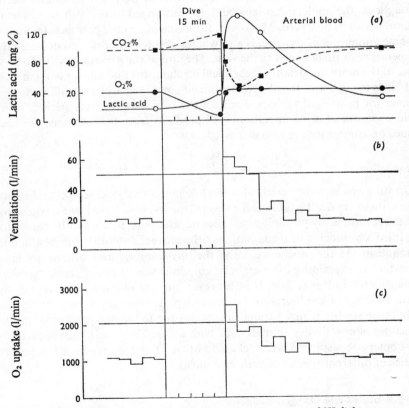

Fig. 1. *Respiratory parameters from a diving experiment in a grey seal (Halichoerus grypus)*
(*a*) Arterial content of lactic acid (○), O_2 (●) and CO_2 (■). (*b*) Ventilation in litre/min. (*c*) Oxygen uptake in litre/min (redrawn from Scholander, 1940).

submerged for from two to four times longer than the period that the stored O_2 would last if the pre-dive metabolic rate was maintained. As early as 1899 this led Richet to propose that aerobic metabolism decreased during a dive. This concept was later documented by Scholander (1940) in the seal (Fig. 1), duck and penguin. It was also realized that the decreased metabolism was a consequence of physiological reflex adjustments (Irving, 1939). A later series of experiments indicated that this mechanism applies equally to mammals, birds, reptiles (Andersen, 1961), frogs (Leiverstad, 1960) and, indeed, fishes (when exposed to air) (Leivestad et al., 1957).

Physiological Reflex Adjustments

The mechanism of outstanding importance for the understanding of the diving animal's ability to sustain submersion for prolonged periods, is its ability to lower profoundly the extraction of oxygen from the blood. This is accomplished by neurogenic transformation of the cardiovascular system into a heart–brain–lung preparation. This transformation is caused by a peripheral vasoconstriction, which virtually stops blood flow in all major systemic circuits except for those of the myocardium, adrenals and the central nervous system. It follows that the total blood oxygen store is utilized by these tissues, whereas other tissues such as muscles and even kidneys (Elsner et al., 1966) have to base their aerobic metabolism on local stores of myoglobin.

In order to balance the increase in peripheral resistance (i.e. to avoid a large increase in pressure) after vasoconstriction, cardiac output is greatly decreased (Elsner et al., 1964; Folkow et al., 1967). This is accomplished by the following adjustments which minimize the myocardial oxygen requirement: the heart rate is greatly decreased (Fig. 2), and myocardial contractility is decreased (Ferrante & Opdyke, 1969; Blix et al., 1975c) without increasing ventricular-wall tension (Blix et al., 1975c) and without decreasing the stroke volume (Elsner et al., 1964). Consistent with this decrease in ventricular work, Blix et al. (1975c) have demonstrated that myocardial blood flow is decreased to only 7–10% of the pre-dive

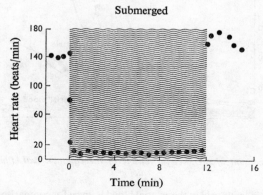

Fig. 2. *Heart rate of a grey seal (H. grypus) before, during and after a dive lasting 12 min*

A. S. Blix (unpublished work).

value. It follows that virtually only the brain, which in spite of a somewhat increased hypoxic tolerance (Elsner *et al.*, 1970) is still most easily damaged by low oxygen tensions, consumes oxygen from the blood reservoir. The physiological reflex mechanisms underlying the cardiovascular transformations have been extensively investigated in the duck by Blix and collaborators (e.g. Blix *et al.*, 1974, 1975*d*; Blix 1975).

In spite of these physiological modifications, and in spite of a marked Bohr shift (i.e. decreased O_2 affinity of haemoglobin at increased H^+ ion concentrations in the blood) (Andersen *et al.*, 1965) making it possible for the animal to endure aerobic energy metabolism for some time, the oxygen stores are limited. The arterial pO_2 will therefore inevitably decrease throughout the dive (Fig. 1) and during prolonged dives values of less than 10 mmHg may be observed. Consequently, anaerobic energy metabolism becomes more and more important even in the brain and the heart.

Biochemical Adaptations to Anaerobiosis

Consistent with the concept of anaerobic metabolism, Scholander *et al.* (1942*a*) observed a marked increase in skeletal-muscle lactate concentration (Fig. 3) during the dive. Further, recent research has shown some biochemical peculiarities which might be judged as adaptations to anaerobic energy metabolism. Kerem *et al.* (1973) have reported that both brain and heart glycogen contents are two to three times greater in the seal than in most other mammals, whereas the skeletal-muscle values are comparable with e.g. those of the cat and the dog. George *et al.* (1971) have also determined the fat content of seal skeletal muscle and found it to be low. On the other hand, A. S. Blix & J. K. Kjekshus (unpublished work) have observed that the free fatty acid concentration of the blood of the seal is comparable with that of humans, and that the free fatty acid concentration decreases and glucose concentration increases during a dive. Remembering that

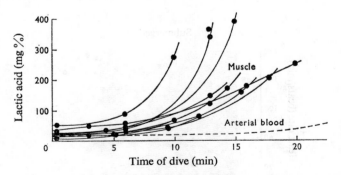

Fig. 3. *Lactic acid content of the skeletal muscle of the common seal* (Phoca vitulina) *during diving*

The average content of lactic acid in the arterial blood is given for comparison (from Scholander *et al.*, 1942*a*).

the blubber deposits are uncirculated while the portal circulation is operating, it is possible that free fatty acids are used together with glucose and lactate in the heart at the beginning of the dive and that they are replaced by pure glucose as anaerobic conditions become more prevalent. A final conclusion, however, must await direct measurements of myocardial arteriovenous differences. Moreover, the low fat content of the skeletal muscles of seals makes it most likely that glucose is also important as a substrate in this tissue.

However, Hochachka et al. (1975) have reported that the concentration of succinate and alanine increased markedly in the blood of the seal after a dive. They interpreted these findings as an indication of anaerobic protein catabolism in muscles during the dive. If this were correct, it might partly explain why skeletal muscles have relatively low concentrations of both glycogen and fat. Alternatively, it might be argued that the rise in blood amino acids is due to protein mobilization in the recovery period after the dive. The significance of this phenomenon could be in rebuilding the glycogen stores. This explanation seems likely since the seal consumes an almost sugar-free diet (see below).

As to the enzyme system, Hellung-Larsen & Andersen (1968) have reported that the proportion of M subunits of the lactate dehydrogenase (EC 1.1.1.27) of human lymphocyte cultures increase on hypoxic exposure. They thereby obtained direct evidence in support of Kaplan & Everse's (1972) theory for different functions for the two types of lactate dehydrogenase, the M type being a pyruvate reductase, functioning in an anaerobic environment and the H protein being a lactate dehydrogenase, functioning under largely aerobic conditions. The lactate dehydrogenase system of seals and eider ducks has been studied by Blix & From (1971) and Blix et al. (1973) and that of beavers by Messelt & Blix (1975). It was found that the enzyme in all tissues from diving animals had a much higher activity than that from sheep, and that the brain and the heart had a significantly higher M subunit content than the same tissues of sheep. The skeletal muscle of both seal and beaver had, on the other hand, a much lower M subunit content than the skeletal muscle of sheep. According to Kaplan & Everse (1972) this might be interpreted as poor adaptation. Two recent studies may provide another explanation. Berg & Blix (1973) found that different cell types in the same tissue (liver) had very different isoenzyme patterns, and Messelt & Blix (1975) demonstrated that whereas the muscle of the sheep is composed almost exclusively of white (anaerobic) fibres and contains M subunits almost entirely, the skeletal muscle of the beaver is composed of both red (aerobic) and white fibres in a proportion comparable with the relative amount of M and H subunits in the muscle. These findings might indicate that the lactate dehydrogenase composition is fibre-specific, and that the fibre composition rather than the lactate dehydrogenase pattern of the muscle as a whole reflects adaptation to a diving habit. It might appear strange that the seal and the beaver should have fewer anaerobic fibres in their skeletal muscle than sheep, but remembering that the H protein functions mainly as a lactate dehydrogenase such fibres will evidently be advantageous in the recovery period after the termination of the dive.

Blix (1971) in a search for specific energy stores has measured the creatine content of different tissues in some diving animals. No significant differences between diving and non-diving animals were found. Considering the decreased work load on the heart during the dive, however, creatine phosphate might well be of importance in this tissue during diving.

Blessing & Hartschen-Niemeyer (1969) have also made a contribution to the understanding of myocardial metabolism in diving mammals. They found that the myoglobin content of seal heart was twice as high as that in man. Their studies were, however, carried out on corpses obtained from zoological gardens, and their values are certainly an underestimate.

I have now cited most of the few papers available on biochemical adaptations to anaerobiosis in diving animals, and only a rather incomplete picture can be obtained from these studies alone. However, Hochachka and co-workers have constructed a more complete scheme by extrapolating much of our knowledge on diving reptiles, on invertebrates and on terrestrial mammals as well. The result (Hochachka & Storey, 1975) is a most challenging suggestion as to the way in which diving animals in general cope with the underwater periods, i.e. how biochemically they are able to oscillate between aerobic and anaerobic metabolism. Their views may be summarized as follows. The basic theme of our analysis of metabolic consequences of diving is that muscle glycolysis, even in the laboratory rat, is already a most impressive anaerobic machine, and further improvement of its capacity and efficiency in muscle of diving vertebrates seems to have involved only a modest number of modifications. Thus the steady-state concentrations of a few glycolytic enzymes are increased, reflecting a higher overall glycolytic potential and an improved capacity to maintain $NAD^+/NADH$ ratios during anoxic stress. To retain control of the higher glycolytic capacity, at least two additional modifications are known: (i) the ratio of fructose diphosphatase to phosphofructokinase in the porpoise (Storey & Hochachka, 1974) is one of the highest thus far reported, the significance being an amplification of the AMP signal for glycolytic activation (Newsholme, 1971), and (ii) muscle pyruvate kinase, although having a lower specific activity, occurs as a regulatory enzyme, which is highly sensitive to feedforward activation by fructose 1,6-diphosphate and feedback inhibition by ATP, alanine and citrate. The fructose 1,6-diphosphate feedforward activation presumably functions during the aerobic–anaerobic transition in the dive, whereas the feedback inhibitions by ATP, alanine and probably citrate (all acting in effect as end products of aerobic fatty acid catabolism) appear to function during the anaerobic–aerobic transition at the end of diving. The latter characteristic emphasizes another important consequence of the diving habit: a metabolic organization that swings between an anaerobic glycogen–based fermentation and an aerobic fat-based oxidative metabolism. The control requirements imposed on muscle by this metabolic organization have led to the appearance of unusually high titres of aspartate aminotransferase and alanine aminotransferase (Owen & Hochachka, 1974). The mitochondrial form of aspartate aminotransferase is designed to 'spark' the tricarboxylate cycle by increasing the availability of oxaloacetate at the same time that acetyl-CoA is being produced by β-oxidation. Alanine aminotransferase regenerates the α-oxoglutarate required for this process and leads to the accumulation of alanine, which play a key role in turning off glycolysis at this time (Hochachka & Storey, 1975).

Biochemical Adaptation to Low Temperature

Owing to the circulatory readjustments mentioned above, the energy metabolism in some tissues has to proceed at low temperatures. This is particularly apparent in the skin and flippers of seals and whales, where the combined effects of circulatory arrest and the cold surrounding water are tissue temperatures of 1–2°C during diving. Feltz & Fay (1966) have, in fact, demonstrated that the temperature in the skin of arctic seals is too low to allow cell proliferation throughout most of the year. On the other hand, they found that seal skin cells had an outstanding ability to survive prolonged exposure to low temperatures. Recently, a few studies have focused attention on these interesting patterns of energy metabolism (i.e. anaerobiosis in combination with very low temperatures in highly organized mammals).

Somero & Johansen (1970) studied the lactate dehydrogenase and pyruvate kinase (EC 2.7.1.40) activity of the heterothermic flipper artery and the homeothermic renal artery of the seal at different temperatures. They found that substrate affinity varied inversely with temperature in both flipper and kidney arteries. Moreover, Behrisch & Percy (1974) found that the substrate affinity of 6-phosphogluconate dehydrogenase (EC 1.1.1.43) from the heterothermic adipose tissue of the seal was unaffected by a decrease in temperature, whereas adipose tissue from deep homoeothermic deposits displayed a large decrease in apparent

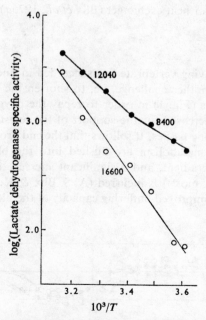

Fig. 4. *Arrhenius plots for ringed seal (*Pusa hispida*) (●) and sheep (○) skin lactate dehydrogenase, showing the differential effect of temperature on the catalytic behaviour of the enzymes*

Lactate dehydrogenase specific activity is given as μmol of NADH oxidized/min per mg of protein. The numbers above each plot represent activation energy (cal/mol) calculated over a given temperature range. Range of temperatures 3.5–42.9°C (from Blix *et al.*, 1975b).

K_m. Blix *et al.* (1975c) have also studied the effect of low temperature on the activity of lactate dehydrogenase from seal and sheep skin. They found that the seal skin lactate dehydrogenase had ten times higher activity at low temperatures than the sheep skin. On the basis of Arrhenius plots of enzyme activity they concluded that this, at least in part, could be explained by the fact that the activation energy of the seal skin enzyme decreased with decreased temperature (Fig. 4), being only half of that for the sheep skin enzyme already at 18°C.

Metabolic Defence Against Hypothermia

The metabolic problems connected with low temperature can be compensated for by mechanisms such as those outlined above. However, the combined effects of decreased general metabolism, peripheral circulatory arrest and the cold surrounding water is a cooling of the body as a whole (Fig. 5) (Scholander *et al.*, 1942b). For animals like the seal, which display an uninterrupted diving pattern throughout the day, this cooling of the body could be limiting to its diving behaviour if effective heating mechanisms were not operating during or immediately after the dive. The presence of large deposits of brown adipose tissue in at least some Pinnipeds is no doubt of importance in this connexion (Blix *et al.*, 1975a). In harp seals the brown adipose tissue embeds the large venous plexuses in the neck (Plate 1) (Hol *et al.*, 1975), on the pericardium and around the kidneys, and we have proposed that this venous plexus–brown fat complex might function as a high-efficiency tubular heat exchanger (Blix *et al.*, 1975a).

The Recovery Period

Once a naturally diving vertebrate has succeeded in undergoing a prolonged dive, it has to face another challenge, i.e. to shorten the recovery period. The animal hyperventilates (Fig. 1) in order to repay the oxygen debt, and cardiac output is markedly increased for the support of the re-established perfusion of the previously ischaemic tissues. It follows that the end products (mainly lactate) of anaerobic energy metabolism are flushed into the blood, because of the great concentration gradient, and a significant increase in arterial osmolarity (from 320 mosM to 375 mosM) is incurred (A. S. Blix & K. Fugelli, unpublished work). In spite of the improved buffering capacity of the blood of diving animals

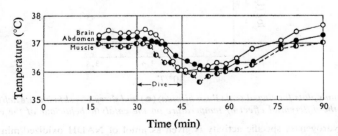

Fig. 5. *Temperatures of brain, abdomen and skeletal muscle of a common seal (P. vitulina) expos to 15 min of submersion in water of 20°C [from Scholander et al. (1940b)]*

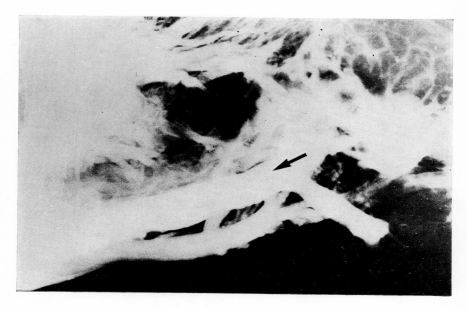

EXPLANATION OF PLATE I

Venous plexus in the neck of a harp seal (*Pagophilus groenlandicus*) *as shown by angiography*

The venous plexus can be seen in the upper left corner, and a part of the skull in the upper right. The horizontal structures (arrow) at the bottom of the picture are the anterior caval veins (from Hol *et al.*, 1975).

A. S. BLIX

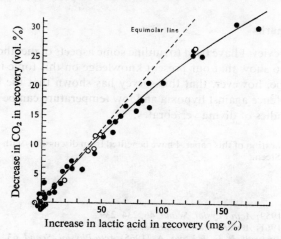

Fig. 6. *Correlation between the increase in lactic acid and the decrease in CO_2 content in the arterial blood of grey (H. grypus) (●) and hooded (Cystophora cristata) (○) seals after the dive*

The results are taken from Scholander (1940). Three experiments are excluded.

(Lenfant *et al.*, 1970), a marked decrease in arterial pH also occurs. It is noteworthy that this arterial hyperosmolarity is eliminated before the rise in arterial lactate has culminated and while the pH is still low (A. S. Blix & I. Stensvold, unpublished work). Scholander (1940) found that arterial CO_2 content decreased just after the dive (Fig. 1), and that the decrease in CO_2 was osmotically equivalent to the increase in arterial lactate (Fig. 6). The most likely, mechanism therefore for osmotic readjustment is respiratory removal of CO_2 from the blood during the period of hyperventilation after the termination of the dive. The reaction can be described by the following equation:

$$CO_2 + H_2O \leftrightharpoons H_2CO_3 \rightleftharpoons H^+ + HCO_3^-$$

The full recovery from a prolonged dive also includes rebuilding the substrate stores, but information on these metabolic events is sparse. Some suggestions, however, seem justified. The anaerobic metabolism has most likely drained the glycogen stores. At least in the seal, these stores are refilled with difficulty, as the diet of such animals is almost free from carbohydrate. Indeed, the milk of the seal is completely devoid of sugar (Sivertsen, 1936), and consequently the seal pup should be considered when principles of mammalian gluconeogenesis are to be investigated. In the adult seal, the glycogen stores can certainly be rebuilt from lactate in the liver, and possibly in the muscles during the recovery period. The high concentration of amino acid in the blood of the seal after the dive (Hochachka *et al.*, 1975) might also indicate that protein is mobilized for the purpose of gluconeogenesis. The possibility of a glyoxalate cycle in the production of carbohydrates has never been investigated. In any case, the ATP necessary for the replenishment of the glycogen stores is most likely derived from oxidation of free fatty acids, which might already have been mobilized in the huge subcutaneous blubber deposits during the dive.

Concluding Remarks

In this short review I have tried to outline some aspects of metabolic adaptation to diving, and to show that our present knowledge on this topic is by no means complete. I hope, however, that this survey has shown that the basic concepts of metabolic defence against hypoxia and low temperature can be obtained from comparative studies of diving vertebrates.

During the preparation of this paper, I have benefited from discussions with Dr. Hans J. Grav and Dr. Johan B. Steen.

References

Andersen, H. T. (1959) *Acta Physiol. Scand.* **46**, 234–239
Andersen, H. T. (1961) *Acta Physiol. Scand.* **53**, 23–45
Andersen, H. T. Hustvedt, B. E. & Løvø, A. (1965) *Acta Physiol. Scand.* **63**, 128–132
Behrisch, H. W. & Percy, J. A. (1974) *Comp. Biochem. Physiol. B* **47**, 437–443
Berg, T. & Blix, A. S. (1973) *Nature (London) New Biol.* **245**, 239–240
Blessing, M. H. & Hartschen-Niemeyer, E. (1969) *Z. Biol.* **116**, 302–313
Blix, A. S. (1971) *Comp. Biochem. Physiol. A* **40**, 805–807
Blix, A. S. (1975) *Acta Physiol. Scand.* **95**, 41–45
Blix, A. S. & From, S. H. (1971) *Comp. Biochem. Physiol. B* **40**, 579–584
Blix, A. S., Messelt, E. B. & From, S. H. (1973) *Comp. Biochem. Physiol. B* **44**, 625–627
Blix, A. S., Gautvik, E. L. & Refsum, H. (1974) *Acta Physiol. Scand.* **90**, 289–296
Blix, A. S., Grav, H. J. & Ronald, K. (1975a) *Acta Physiol. Scand.* **94**, 133–135
Blix, A. S., Messelt, E. B. & Grav, H. J. (1975b) *Acta Physiol. Scand.* **95**, 77–82
Blix, A. S., Kjekshus, J. K., Enge, I. & Bergan, A. (1975c) *Acta Physiol. Scand.* in the press
Blix, A. S., Wennergren, G. & Folkow, B. (1975d) *Acta Physiol. Scand.* in the press
Elsner, R., Franklin, D. L. & Van Citters, R. L. (1964) *Nature (London)* **202**, 809–810
Elsner, R., Franklin, D. L., Van Citters, R. L. & Kenney, D. W. (1966) *Science* **153**, 941–949
Elsner, R., Shurley, J. T., Hammond, D. D. & Brooks, R. E. (1970) *Comp. Biochem. Physiol.* **9**, 287–297
Feltz, E. T. & Fay, F. H. (1966) *Cryobiology* **3**, 261–264
Ferrante, F. L. & Opdyke, D. F. (1969) *J. Appl. Physiol.* **26**, 561–570
Folkow, B., Nilsson, N. J. & Yonce, R. L. (1967) *Acta Physiol. Scand.* **70**, 347–361
George, J. C., Vallyathan, N. V. & Ronald, K. (1971) *Can. J. Zool.* **49**, 25–30
Guard, C. L. & Murrish, D. E. (1973) *Antarctic J.* **8**, 198–199
Hellung-Larsen, P. & Andersen, V. (1968) *Exp. Cell. Res.* **50**, 286–292
Hochachka, P. W. & Storey, K. B. (1975) *Science* **187**, 613–621
Hochachka, P. W., Owen, T. G., Allen, J. F. & Whittow, G. C. (1975) *Comp. Biochem. Physiol. B* **50**, 17–22
Hol, R., Blix, A. S. & Myhre, H. O. (1975) *Rapp. P.V. Reun. Cons. Int. Explor. Mer.* in the press
Irving, L. (1939) *Physiol. Rev.* **19**, 112–134
Kaplan, N. O. & Everse, J. (1972) *Adv. Enzyme Regul.* **10**, 323–336
Kerem, D., Hammond, D. D. & Elsner, R. (1973) *Comp. Biochem. Physiol. A* **45**, 731–736
Leivestad, H. (1960) *Arbok Univ. Bergen. Mat. Nat. Urvitensk, Ser. no.* 4, 1–15
Leivestad, H., Andersen, H. & Scholander, P. F. (1957) *Science* **126**, 505
Lenfant, C., Johansen, K. & Torrance, J. D. (1970) *Resp. Physiol.* **9**, 277–286
Messelt, E. B. & Blix, A. S. (1975) *Comp. Biochem. Physiol.* in the press
Newsholme, E. A. (1971) *Cardiology* **56**, 22–34
Owen, T. G. & Hochachka, P. W. (1974) *Biochem. J.* **143**, 541–553
Richet, C. (1899) *J. Physiol. Pathol. Gen.* **1**, 641–650
Robinson, D. D. (1939) *Science* **90**, 276–277
Scholander, P. F. (1940) *Hvalradets Skr.* **22**, 1–131
Scholander, P. F., Irving, L. & Grinnell, S. W. (1942a) *J. Biol. Chem.* **142**, 431–440
Scholander, P. F., Irving, L. & Grinnell, S. W. (1942b) *J. Cell. Comp. Physiol.* **19**, 67–78
Sivertsen, E. (1936) *Nytt Mag. Nat. Vid.* **75**, 183–185
Somero, G. N. & Johansen, K. (1970) *Comp. Biochem. Physiol.* **34**, 131–136
Storey, K. B. & Hochachka, P. W. (1974) *Comp. Biochem. Physiol. B* **49**, 119–128

Biochem. Soc. Symp. (1976) **41**, 179–204
Printed in Great Britain

Adaptations with Respect to Salinity*

By E. SCHOFFENIELS

Laboratoire de Biochimie Générale et Comparée, Université de Liège,
17 Place Delcour, B-4020 Liège, Belgium

Synopsis

Amino acids contribute up to about 50% of the intracellular osmotic pressure of aquatic invertebrates. Since their concentration varies according to the salinity of the medium (high in sea water, low in fresh water) euryhaline invertebrates are good models for studying the mechanisms involved in the control of amino acid metabolism. During hyperosmotic stress CO_2 production and O_2 consumption decrease whereas the reverse is true when the animal is submitted to a hypo-osmotic stress. Nitrogen excretion (as NH_3) increases in media of low salinity and the concentration of cyclic AMP increases during hyperosmotic stress. Moreover, blood proteins and haemocyanin are more concentrated in individuals adapted to media of low salinity. To explain the situation, three main mechanisms can be considered: (*a*) hydrolysis and synthesis of blood proteins; (*b*) transport of amino acids across the cell membrane; (*c*) control of the turnover rate of some amino acids. Results obtained on whole animals as well as on isolated tissues indicate that some amino acids are released from the cells and carried via the haemolymph to the posterior pairs of gills where they are oxidized (mechanism *b*) or to an organ (hepatopancreas?) where they are used for blood protein synthesis (mechanism *a*). The use of labelled substrates demonstrates that the turnover rate of amino acids is controlled by the salinity of the environment (mechanism *c*). It is suggested that inorganic ions trigger the metabolic response by directing reducing equivalents toward oxygen or 2-oxo acids through control of the catalytic activity of dehydrogenases.

Introduction

From experimental as well as from theoretical considerations it is obvious that the relation existing between the activity of water inside and outside a cell cannot be maintained by a mere transfer of water across the plasma membrane, thus implying as the case may be an increase or decrease in cell volume. Indeed recent progress in biochemistry as well as in electron microscopy favours the idea that the cell interior is highly organized and that the macromolecules surrounded by the intracellular fluid form a system closer to the physicochemical state of a gel than to a true solution. Inflow of extracellular solution tends to produce a swelling of the cell due to a Gibbs–Donnan effect. For instance if erythrocytes were at Gibbs–Donnan equilibrium the osmolarity of the cytoplasm would be expected to exceed that of the medium by 25–30 mosm and the excess of hydrostatic pressure

* Dedicated to Professor Marcel Florkin on the occasion of his 75th birthday.

to be of the order of 5800 mm of water (Tosteston, 1964). This last value is obviously much too large to be compatible with the mechanical resistance of the cellular membrane. From the values computed by Rand & Burton (1964), we know that the pressure inside the erythrocyte is 2.3 ± 0.8 ml of water higher than that outside, a value that agrees well with the estimate of Cole (1932) for the internal pressure of the sea-urchin egg. This indicates that the animal cells are not in Gibbs–Donnan equilibrium and that they resort to specialized mechanisms to avoid bursting (Schoffeniels, 1967).

Swelling can be avoided if the cell possesses rigid membranes or a membrane impermeable to water and to a large fraction of the solutes present in the extracellular medium. The first solution is found in plant cells and some bacteria (Rothstein, 1959). The existence of tough walls that resist the large hydrostatic pressure difference balancing the osmotic inflow of solution thus offers an interesting way of explaining why in the course of evolution the biosynthesis of cellulose and some mucopeptides occurred. However, in forming these walls the plant cells have had to give up motility.

The second solution, impermeability of the cell, has never been shown to occur. On the contrary, the use of isotopic tracers has demonstrated that the water as well as the inorganic ions contained in a living system can traverse the membrane. Despite these facts, in normal conditions of metabolism no swelling of animal cells occurs. But as soon as the metabolic activity of the cell is impaired, cell swelling occurs. The phenomenon has been extensively studied by various authors (Opie, 1949; Mudge, 1951; Deyrup, 1953; Leaf, 1956; Whittam, 1956; Robinson, 1960). Some of them consider that the swelling of various tissues under conditions of metabolic inhibition is indicative of a gradient of water activity between the cell interior and the extracellular medium. Without going into a detailed presentation of a long debate that has involved many scientists for the last 50 years, it is sufficient to say that today it is generally accepted that the osmotic pressure of a cell interior is very close to that of the extracellular fluid. The extremely high permeability to water of cellular membranes and the 'smallness' of the cells also favour the absence of a gradient of water activity between cytoplasm and extracellular medium. Any gradient of activities that could develop would be rapidly dissipated in considerably less than 1 s in most cells. Thus water traverses the membrane more rapidly than do virtually all polar solutes and very much more rapidly than ions. The water activity inside the cell may thus be related to the metabolic activity of the membrane as well as to that of the gel-like network of polymers constituting the integrated metabolic sequences.

It is obvious that any impairment of the metabolic activity of the cells as a whole should lead to a swelling not necesarily because of a difference in osmotic pressure, but because of the existence of polymer chains that (a) may accept water as solvent, (b) contain ionized groups and (c) control together with the cellular membrane the activity of the solutes within the cell. In this respect the active transport of cations must be considered, as discussed elsewhere (Schoffeniels, 1967), to be of prime importance in the maintenance of the cell volume. The cell is thus in a state far from thermodynamic equilibrium and energy must be fed into the system to keep it in this situation. The swelling observed under conditions of metabolic impairment is thus an illustration of this fact.

Table 1. *Free amino acid content of serum, muscle and nerve from Homarus vulgaris and Astacus fluviatilis*

Values are expressed in mmol/kg wet weight or in mmol/litre of serum. —, Not estimated.

Amino acid	H. vulgaris			A. fluviatilis		
	Serum*	Muscle*	Nerve†	Serum‡	Muscle‡	Nerve†
Alanine	0.98	23.0	11.33	1.14	6.14	2.39
Arginine	0.01	60.5	4.88	0.20	54.35	—
Aspartate	0.50	2.0	32.31	0.17	7.33	5.10
Cystine	—	—	0.84	—	—	—
Glutamate	0.20	34.1	9.76	2.02	28.91	2.00
Glycine	3.2	202.0	20.55	0.80	33.73	1.45
Histidine	0.2	1.0	0.23	0.07	2.25	—
Isoleucine	—	—	0.53	0.45	1.35	0.21
Leucine	0.3	6.0	0.33	0.22	2.48	0.20
Lysine	0.15	3.2	0.36	0.20	6.98	—
Methionine	0	1.0	—	—	—	—
Phenylalanine	Traces	1.0	—	0.06	1.26	0.23
Proline	0.5	104.0	13.89	0.28	10.78	0.64
Serine	—	—	6.57	—	—	2.41
Taurine	—	—	16.13	—	—	0.47
Threonine	0	1.0	1.72	0.27	2.73	0.45
Tyrosine	0.2	0	—	0.09	1.06	0.26
Valine	0	3.0	—	0.51	2.16	0.32

* Camien *et al.* (1951).
† Gilles & Schoffeniels (1968*a*).
‡ Duchâteau & Florkin (1961).

A particularly interesting situation is offered by the case of the so-called euryhaline species, i.e. those marine species that are able to invade brackish or fresh water or the freshwater species that are able to invade salted water of various concentrations. In the case of the aquatic invertebrates it has been shown that the osmotic pressure of the blood changes with the variation of the environmental salinity. Since the cell volume remains constant whatever the change in osmotic pressure observed in the blood, we have to conclude, from the above considerations, that the water activity of the cell interior is controlled by the concentration of intracellular osmotic effectors in such a way as to keep the equality between the osmotic pressure of the cell and that of its surrounding extracellular fluid. We have known for the last 25 years that the amino acid concentration in the cells of marine invertebrates is higher than that found in freshwater species (Table 1). This observation strongly suggested that amino acids could well be important osmotic effectors in the cell-volume regulation. This interpretation was shown to be correct during the study of the amino acid content of muscle isolated from euryhaline species adapted to media of various salinities. Table 2 gives the variation of the intracellular pool of free amino acid in a euryhaline species *Eriocheir sinensis* Milne Edwards living in media of various concentrations. In this experiment the inorganic composition of the intracellular fluid has also been determined in order to establish the osmotic balance. Fig. 1 gives the results obtained with nerve fibres isolated from the meropodite of the claws and walking legs of the same species (Schoffeniels, 1964*a*, 1967). It can be seen that the intracellular composition of the nerve fibres is also subjected to variations. As the hydration of the muscle and of the nerve fibres is approxim-

Table 2. *Intracellular osmotic effectors in muscle of E. sinensis adapted to fresh water and to sea water*

The results were taken from Bricteux-Grégoire *et al.* (1962).

Intracellular osmotic effectors	Content (mosm/litre of water) in			
	Fresh water	Sea water	Fresh water	Sea water
Cl⁻	76.0	153.1	44.6	166.9
Na⁺	68.5	140.8	41.4	146.9
K⁺	56.8	159.0	84.5	133.1
Ca²⁺	11.7	8.1	5.2	11.2
Mg²⁺	9.2	22.4	9.2	25.3
Total inorganic effectors	222.2	483.4	184.9	483.4
Alanine	17.1	46.1	18.1	71.9
Arginine	36.7	56.0	36.5	54.7
Aspartic acid (total)	5.4	12.2	3.6	11.7
Glutamic acid (total)	15.0	36.8	10.3	28.2
Glycine	46.5	73.4	57.0	108.5
Isoleucine	1.4	4.6	1.0	3.2
Leucine	2.2	6.1	1.7	5.4
Lysine+histidine+X	9.6	21.7	14.3	18.5
Phenylalanine	0.0	tr.	0.0	tr.
Proline	18.2	37.3	4.7	23.7
Serine	5.2	7.6	2.6	6.3
Threonine	4.4	17.2	4.4	15.3
Tyrosine	0.0	tr.	0.0	tr.
Valine	0.0	8.1	0.0	6.9
Total amino acids determined	161.7	327.0	154.2	354.3
Taurine	14.1	13.6	20.5	27.7
Trimethylamine oxide	49.9	73.9	45.3	75.8
Betaine	9.5	6.9	25.7	21.0
Undetermined nitrogen	108.5	187.7	89.3	131.9
Total effectors determined	565.9	1092.5	520.0	1094.1
Calculated osmolar concentration $(\Delta/1.87) \times 1000$	588.0	1117.6	588.0	1117.6

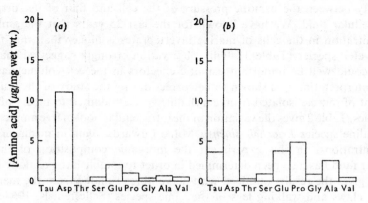

Fig. 1. *Free amino acids in nerve fibres of E. sinensis Milne Edwards*

Free amino acids in nerve fibres isolated from animals kept in fresh water (*a*) and sea water (*b*).

ately the same in sea water and in fresh water the reversible variation of the amino acid content could only depend on active modification (Florkin, 1956). As the same kind of difference was found in the concentrations of free amino acids in *Carcinus maenas* living in sea water or in brackish water, it was proposed by Florkin to consider the variation of the amino acid content resulting from the change in concentration in the medium 'as accomplishing an intracellular regulation acting against the water movement between cells and body fluid as a consequence of change in concentration in the latter' (Duchâteau & Florkin, 1956). The term iso-osmotic intracellular regulation has been coined by Duchâteau & Florkin (1956) to define the regulation which leads to the active adjustment of the intracellular osmotic pressure to the new osmotic pressure of the body fluid, more or less presenting a change of hydration of the cells. This term replaces by a definite molecular concept the vague notion of cellular adaptation and also introduces the distinction between the iso-osmotic intracellular regulation and what is known as the osmotic regulation of body fluids, which is an aniso-osmotic regulation, keeping the concentration of body fluids above that of the external medium in freshwater forms, below in marine bony fishes or equal in marine invertebrates. Since the existence of the iso-osmotic cellular regulation was demonstrated in *E. sinensis* by Duchâteau & Florkin (1955), the phenomenon has been observed repeatedly in many laboratories and was shown to occur as a general phenomenon in many aquatic invertebrates [for a review see Florkin & Schoffeniels (1969) and Schoffeniels (1973, 1974, 1976)].

As shown in Table 2 the inorganic ions form an important part (40%) of the osmotically active constituent in *E. sinensis*. The rest of the osmotic pressure is accounted for approximately by the presence of small organic molecules, among which the amino acids determined constitute more than 50%.

The contribution of trimethylamine oxide to the sum of osmotically active compounds is important and quantitatively similar to that of glycine. Taurine and betaine contribute to about the same degree as alanine and proline. The contribution of the sum of the undetermined nitrogen constituents amounts to approximately twice that of trimethylamine oxide or glycine. In spite of the fact that the osmotic coefficient of the constituents determined cannot be considered as equal to unity, it appears that the nitrogen constituents added to the inorganic ones make up the greatest part of the osmotically active components, if not the whole.

Thus when the animals are adapted to sea water, an increase is observed in the concentrations of inorganic constituents, amino acids, trimethylamine oxide and the undetermined nitrogen. The sum of these osmotically active constituents corresponds to the greatest part, or even the total, of the osmotic components responsible for the new equilibrium which is accomplished between the cell component and the blood plasma, the concentration of which becomes equal to that of sea water.

Two interesting observations can be made when considering the amino acid pattern of muscle and nerve fibres. It is obvious that arginine is more concentrated in muscle than in nerve whereas the reverse is true as far as aspartic acid is concerned. They may be considered as a biochemical characteristic of the type of cellular differentiation considered. On the other hand, it is also apparent that

7

Table 3. *Amino acid concentration in muscles of several crustaceans adapted to media or various salinities*

Values are given in μmol/100mg wet weight. tr. indicates trace; dashes indicate that values were not determined. SW, Sea water; FW, fresh water; SW/2, 50% sea water; SW/3, 30% sea water. Histidine is measured together with lysine for *Eriocheir sinensis*. The most important osmotic effectors (quantitatively) are underlined.

	Eriocheir sinensis*		Carcinus maenas†		Leander serratus‡		Astacus astacus§		Callinectes sapidus‖	
	SW	FW	SW	SW/2	SW	SW/3	SW/2	FW	SW	SW/2
Alanine	3.37	1.39	2.05	0.96	1.69	0.26	2.11	1.32	2.05	1.24
Arginine	4.13	2.99	3.62	3.37	2.58	2.07	4.42	3.70	7.44	5.24
Aspartic acid	0.86	0.29	0.39	0.27	0.18	0.05	0.95	0.80	0.42	0.17
Glutamic acid	2.11	0.84	3.60	1.71	0.32	0.11	5.05	3.43	0.55	0.19
Glycine	8.00	4.64	10.07	7.13	11.2	9.5	2.43	2.09	19.75	15.50
Histidine	—	—	0.01	0.004	0	0	0.16	0.06	0.13	0.04
Isoleucine	0.24	0.08	0.17	0.03	0.21	tr.	0.34	0.17	0.1	0.09
Leucine	0.40	0.14	0.26	0.05	0.34	tr.	0.40	0.21	0.19	0.19
Lysine	1.38	1.16	0.19	0.10	0.16	0.05	0.71	0.56	0.19	0.005
Phenylalanine	tr.	0	0.05	0.012	tr.	0	0.06	0.09	0.23	0.038
Proline	3.50	0.77	9.84	1.83	2.17	0.64	2.57	0.78	4.04	2.68
Serine	0.47	0.21	—	—	0.55	0.12	—	—	2.86	0.28
Taurine	2.06	1.67	—	—	2.56	2.32	—	—	3.79	2.05
Threonine	1.14	0.36	0.33	0	0.21	tr.	0.56	0.27	—	—
Tyrosine	tr.	0	0.03	0.006	tr.	0	0.10	0.12	—	—
Valine	0.50	0	0.33	0.046	0.37	tr.	0.46	0.11	0.29	0.30
Total	28.16	14.54	30.9	15.5	22.5	15.12	20.32	13.71	42.03	28.013

* Bricteux-Grégoire *et al.* (1962).
† Duchâteau *et al.* (1959).
‡ Jeuniaux *et al.* (1961).
§ Duchâteau & Florkin (1961).
‖ Gilles (1974a).

alanine, glycine and proline are quantitatively the most important amino acids involved in the control of the intracellular activity of water in the case of the muscle, whereas aspartic acid, proline, glutamic acid, alanine, serine and threonine are involved in the cell-volume regulation of the nerves. There are species differences. As can be seen in Table 3 serine is obviously of less quantitative importance in the iso-osmotic intracellular regulation of the muscle fibres of *E. sinensis* for instance.

In Table 3, I have indicated the amino acids that are quantitatively the most important osmotic effectors in the muscle of Crustacea. This will help us to define the metabolic sequences that could be relevant to our biochemical approach to the problem.

Besides the change in nitrogen metabolism noted above, adaptation with respect to salinity involves other metabolic modifications.

Table 4. *Osmotic stress and oxygen consumption of isolated tissues of two euryhaline Crustacea*

The results are expressed as μl of O_2/h per mg dry weight; 10 mM-glucose was added and the O_2 consumption measured 60 min after the stress. The results are those of Gilles (1974a).

	Hypo-osmotic stress		Hyperosmotic stress	
	Control	Experiment	Control	Experiment
Callinectes sapidus				
Nerve bundle	5.56	7.34	4.06	3.34
Muscle	4.95	6.20	3.15	2.10
Eriocheir sinensis				
Nerve bundle	2.46	3.35	4.04	2.92

The oxygen consumption in the euryhaline species is increased on transfer from sea water to diluted sea water (Gilles & Schoffeniels, 1965; King, 1965). This is also true in isolated tissues as shown by considering the results of Table 4 in which the oxygen consumption is recorded when muscle fibres and isolated bundles of axons from euryhaline Crustacea are submitted to osmotic stress. When the O_2 consumption decreases, CO_2 production also decreases and vice versa.

Osmotic adaptation is also paralleled by a modification of nitrogen excretion, which increases when a euryhaline species is transferred to diluted medium and decreases during the adaptation to a hyperosmotic medium. This has been demonstrated for *Carcinus maenas* (Needham, 1957) and *E. sinensis* (Florkin *et al.*, 1964). From the above facts it is likely that the regulation of the intracellular amino acid pool may depend on the acquisition of mechanisms controlling (*a*) the turnover rate of some proteins, (*b*) the permeability of the cell membrane towards amino acids and (*c*) the relative rate of anabolism and catabolism of amino acids.

In the case of muscle cells, the first possibility can certainly be ruled out. By measuring the variation in free amino acids before and after hydrolysis of proteins, an indication may be obtained as to the possible participation of intracellular protein in the cell adaptation to an increase in the osmotic pressure of the blood. Table 5 shows that when *E. sinensis* is adapted to sea water, the increase in free alanine is accompanied by a parallel increase in the total alanine. The results also show that although the free proline has markedly increased in the muscle, the amount of total proline has increased proportionally, which indicates that the proline obtained from the hydrolysis of protein did not vary significantly. These

Table 5. *Comparison of free and total alanine and proline in the muscle of E. sinensis adapted to fresh water and sea water*

The results show the adaptation after 6 days (Florkin *et al.*, 1964).

	Free alanine (mg/100 g of dry muscle)			Total alanine (mg/100 g of dry muscle)			Free proline (mg/100 g of dry muscle)			Total proline (mg/100 g of dry muscle)		
Crab number	Fresh water	Sea water	Variation	Fresh water	Sea water	Variation	Fresh water	Sea water	Variation	Fresh water	Sea water	Variation
1	1764	2899	+1135	5582	6777	+1195	1966	2575	+609	4451	4614	+163
2	—	—	—	—	—	—	953	1755	+802	3783	4531	+748
3	1794	3183	+1389	5633	6594	+961	718	1446	+728	3707	4569	+862

values are in agreement with the conception of a net entry of alanine and proline into the intracellular pool during the adaptation to sea water rather than a liberation from some tissue reserve. We have also examined the protein composition of *Eriocheir* muscle and have been unable to detect any change in the electrophoretic pattern of the protein during the adaptation of *E. sinensis* to sea water. Therefore mechanisms (*b*) and (*c*) seem to be the only ones at play in the iso-osmotic intracellular regulation. It is thus shown in isolated nerves as well as in intact animals that during a hypo-osmotic shock amino acids leak out of the cell into the blood (Gilles & Schoffeniels, 1969*a*; Vincent-Marique & Gilles, 1970). This is particularly apparent when proline is considered. Since little or no amino acids are excreted into the surrounding medium by the intact animal one has to assume that the amino acids and more specifically proline are oxidized at the level of the excretory organs.

If on the basis of our experimental results it is reasonable to exclude the participation of intracellular proteins in the control of the intracellular amino acid pool this does not seem to be the case for blood. It can indeed be shown that the amount of proteins in the blood varies according to the salinity of the environment of the animal as shown by the unpublished results of R. Gilles obtained in my laboratory (Table 6).

Haemocyanin is the O_2 carrier in the blood of Crustacea: it represents around 80% of the total amount of proteins in the plasma. It is thus interesting to know if the large variations observed during osmotic stress are mainly due to a change in the haemocyanin concentration. It would certainly make sense to have more haemocyanin when the animal is in fresh water since the O_2 requirements of the tissue are much higher. On the other hand, since the molecular weight of the haemocyanin in Arthropods varies between 300000 and 500000, this could be a means of storing amino acids in a form with little thermodynamic activity. In other words the amino acids released by muscles and nerve fibres could be used (maybe in the hepatopancreas) for the synthesis of blood protein including haemocyanin. The blood would then be the storage place for amino acids when the animal is transferred to a dilute medium. On return to a concentrated medium the proteins would be hydrolysed and the amino acids crossing the cell membrane could act as an osmotic effector in the intracellular fluid. This interpretation, though in agreement with the facts so far at hand, needs further experimental evidence.

Since it is generally accepted that the copper content of the blood of Molluscs and Crustacea is related to the concentration of haemocyanin, the copper concentration has been determined in the plasma of *E. sinensis* adapted to sea

Table 6. *Protein content of the plasma of euryhaline crabs adapted to media of various salinities*

The results are expressed in g/litre of plasma, and are means ±S.E.M.

Protein content (g/litre of plasma)

	Sea water	50% Sea water	Fresh water
E. sinensis	34.6 ± 8.12 (*n* = 9)	—	65.3 ± 14.2 (*n* = 17)
C. maenas	27.7 ± 14.0 (*n* = 36)	42.6 ± 23.45 (*n* = 12)	—
A. fluviatilis	—	33.2 ± 5.8 (*n* = 10)	50.5 ± 12.6 (*n* = 6)

Table 7. *Copper content of proteins obtained by* $(NH_4)_2SO_4$ *precipitation from plasma of E. sinensis adapted to sea water and fresh water*

Results are expressed as μg/ml of plasma.

Sea water	Fresh water
17.5	35.5

water or to fresh water. After coagulation of the blood pooled from several animals, the plasma is collected and the proteins are precipitated by 25–50%-satd. $(NH_4)_2SO_4$. The precipitate is dissolved in 20mM-Tris/HCl, pH 7.4, and the solution used for electrophoretic analysis or protein determination (see Table 6). The copper content is determined on a sample by atomic absorption analysis. The results of Table 7 show that when the animal is adapted to fresh water the copper content is twice as high as when it is adapted to sea water, a result which agrees with the above hypothesis that more haemocyanin would be found in the situation where the metabolism of the animal is geared to aerobiosis. It also fits in with the idea that some amino acids released from the cells are stored as blood proteins with low thermodynamic activity.

To test the third possibility, i.e. the control over the synthesis and catabolism of amino acids, various substrates labelled with ^{14}C have been used. The experiments are performed on the isolated ventral nerve chain of the lobster *Homarus vulgaris* and the crayfish *Astacus fluviatilis* (Gilles & Schoffeniels, 1969c). Nerve bundles isolated from the walking legs of *Callinectes sapidus* and *E. sinensis* are also utilized. The preparation is incubated in an appropriate saline solution and either submitted or not to osmotic stress. Two main lines of experiments are performed. After an incubation period of up to 3h the amino acids are separated and their labelling with ^{14}C investigated. In another set of experiments, the $^{14}CO_2$ production is measured to assess the possibilities of oxidation of a given substrate by the tissue under investigation. The viability of the preparation is estimated by measuring the response to an electrical stimulation. The main conclusions that can be drawn from these studies are as follows (see also Florkin & Schoffeniels, 1969; Schoffeniels, 1973, 1975; Gilles, 1974a,b). Alanine, glutamate, aspartate, serine, glycine and taurine are labelled if [U-^{14}C]glucose and [1-^{14}C]pyruvate are substrates in the case of the lobster nerve chain.

The hydrocarbon skeletons of the labelled amino acids are thus derived either from pyruvate (alanine, serine and glycine) or intermediates of the tricarboxylate cycle (glutamate and aspartate). Moreover the values of the specific radioactivities are in the order foreseen by the metabolic sequences involved. The specific radioactivity of taurine is very close to that of alanine. Since cysteine is never labelled under our experimental conditions, we have to assume that a still unknown pathway is operative in taurine biosynthesis.

With glucose as substrate it is found that the specific radioactivities of glutamate and aspartate are close to each other, a fact that may be interpreted as indicating either a compartmentation of 2-oxoglutarate or that a route other than the oxidation of succinate derived from 2-[^{14}C]oxoglutarate is responsible for the production of ^{14}C-labelled oxaloacetate. Carboxylation of pyruvate is the most obvious candidate since aspartate, taurine and glutamate are labelled when $H^{14}CO_3^-$ is used as substrate (see also Cheng & Mela, 1966).

It is interesting to notice that with [U-^{14}C]arabinose, taurine, glutamate and aspartate are labelled and the specific radioactivities are in the order indicated (Gilles & Schoffeniels, 1969b). Since arabinose may be derived from xylulose, it is tempting to assume that in Crustacea, the pentose phosphate pathway may provide 2-oxoglutarate outside the operation of the tricarboxylate cycle. This interpretation is only correct if one may exclude production of ^{14}CO$_2$ from arabinose, which could then be used for the carboxylation of pyruvate.

These experiments have been repeated on isolated axons of C. sapidus and E. sinensis. [1-^{14}C]Glycerate and H^{14}CO$_3^-$ together with [U-^{14}C]glucose (or [1-^{14}C]glucose or [6-^{14}C]glucose) have been used as substrate. Alanine, aspartate and glutamate are always labelled. Serine and glycine are found to be radioactive in some experiments, but their specific radioactivities are always low as is the case with the lobster. Radioactivity is never detected in proline despite the fact that glutamate is always labelled.

The above experiments have been performed on nerve bundles submitted to osmotic stress and the pattern observed followed that of the intact animals. However, in the case of a hyperosmotic stress, part of the increase in amino acid concentration has to be explained by a shrinkage of the axons amounting to up to 25%. If a net synthesis of alanine for instance is obvious, in the case of proline and glycine, most of the concentration variation may be explained by the volume change. This may thus explain the low level of labelling actually found.

As discussed below, 3':5'-cyclic AMP seems to be necessary for the isolated tissue to regulate its volume correctly. If one assumes (see below) that cyclic AMP stimulates the catalytic activity of the phosphorylase, thus providing more glucose and consequently more oxo acids, this may explain why little or no radioactivity is found in glycine or proline in an isolated nerve bundle since the large amounts of oxo acids necessary in those conditions are not available. It may also well be that glycine and proline are compartmentalized and only accessible to rapid metabolic turnover in very precise circumstances requiring cyclic AMP. We have, however, to investigate this problem further before proposing a more definite explanation.

The oxidation of various substrates has been observed by measuring ^{14}CO$_2$ production. Thus in the case of C. sapidus, E. sinensis, A. fluviatilis and H. vulgaris, ^{14}CO$_2$ evolution is observed with the following radioactive substrates: alanine, aspartate, leucine, arginine, glutamate, serine, [U-^{14}C]glucose, [1-^{14}C]-glucose, [6-^{14}C]glucose, [1-^{14}C]pyruvate and [1-^{14}C]acetate.

During the hypo-osmotic stress the amount of ^{14}CO$_2$ produced increases, whereas the converse is true when hyperosmotic stress is applied. This is a good indication that a control over the turnover rate of amino acids does exist in the euryhaline Crustacea (see also Huggins & Munday, 1968).

Thus the adaptation to medium of various salinities involves a large number of mechanisms. At the cellular level it is likely that the regulation of the intracellular amino acid pool depends on the acquisition of mechanisms controlling the relative rate of anabolism and catabolism of the amino acids concerned. It is also evident that the permeability of the cellular membrane to amino acids changes during an osmotic stress.

The oxygen consumption as well as the CO$_2$ production measurements favour

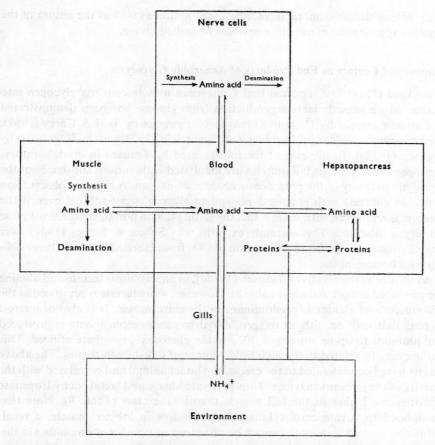

Fig. 2. *Main aspects of nitrogen metabolism involved in the iso-osmotic intracellular regulation in euryhaline Crustacea*

the idea that the reducing equivalents are directed either toward oxygen or toward oxo acids depending on the type of stress imposed on the cell. The protein metabolism seems also to be directly involved in the osmotic regulation of the animal. Since the concentration of proteins is lower in the blood of an animal adapted to a concentrated medium, it is tempting to assume that some proteins have been hydrolysed to give up at least some of the amino acids that are necessary for the iso-osmotic intracellular regulation. Fig. 2 summarizes the main features of the adaptation of Crustacea to media of various salinities. In muscle and nerve cells the transport of amino acids across the cell membrane controls, together with the biosynthesis and the oxidation rates, the intracellular pool of free amino acids. The amino acids crossing the cell membrane during a hypo-osmotic shock are either oxidized (e.g. proline) in the gills or used in blood protein synthesis, including haemocyanin. This operation could well take place in the hepatopancreas.

We shall now examine in more detail the metabolism of the amino acids playing a key role in the iso-osmotic intracellular regulation as well as the nature of the signals triggering the metabolic responses described above.

Alanine and Lactate as End Products of Anaerobic Glycolysis

Boyland (1928) first reported that Crustacean muscle converts glycogen into lactate. More recently lactate production from glucose has been demonstrated in *Carcinus maenas* by Huggins (1966) in *Uca pugnax* by Teal & Carey (1967), *Pachygrapsus crassipes* by Dendinger & Schatzlein (1973) and in *H. vulgaris* by Trausch (1975a). In the case of the lobster used by Trausch in my laboratory, homogenates of claw or tail muscles are incubated with various substrates of the glycolytic pathway in the presence or absence of oxygen. A relevant observation is that in contrast with glucose 6-phosphate glucose is not utilized even if the medium is reinforced with ATP, a fact suggesting that in homogenates hexokinase activity is inhibited. This certainly explains why Scheer & Scheer (1951) were unable to show a significant production of CO_2 from labelled glucose when working with homogenates.

As shown by the results of Trausch (1976a), in anaerobiosis lactate and alanine are produced in both muscles (Table 8). However, more lactate is produced in the tail muscle and alanine is predominant in the claw muscle. It is also of interest to note that with or without oxygen, dihydroxyacetone phosphate is produced and amounts to up to more than 30% of the glucose 6-phosphate utilized. This finding may be related to the high lipid content of Crustacean tissues. The above results have been extended to the case of the intact animal and correlated with the activity of the relevant enzymes. Thus pyruvate kinase and lactate dehydrogenase activities are higher in the tail muscle than in the claw (Table 9). Note that phosphoenolpyruvate carboxykinase is rather low in lobster muscle, a result indicating that oxaloacetate cannot be produced in significant amounts via this route. Since both malate dehydrogenases (EC 1.1.1.37 and EC 1.1.1.40) are present (Schoffeniels & Gilles, 1970) oxaloacetate can be produced from pyruvate.

As indicated by the results of Table 3 alanine is quantitatively an important osmotic effector. Under the influence of a hyperosmotic stress, the metabolism estimated by the oxygen consumption of the whole animal or of isolated tissues is geared to anaerobiosis. The scheme of Fig. 3 shows how anaerobic glycolysis proceeds with lactate and alanine as end products. The reducing equivalents

Table 8. *Products of anaerobic glucose metabolism and concentrations of related enzyme in the lobster (H. vulgaris)*

The results are means ± s.e.m. They are taken from Trausch (1976a).

Tissues	L-Lactate (nmol/mg of protein)	L-Alanine (nmol/mg of protein)	Lactate dehydrogenase (μmol/min per mg of protein)	Alanine aminotransferase (μmol/min per mg of protein)
Tail muscle	253 ± 55 (n = 17)	174 ± 42 (n = 8)	1.59 ± 5 (n = 18)	0.0091 ± 0.0059 (n = 6)
Claw muscle	151 ± 50 (n = 17)	311 ± 117 (n = 8)	0.27 ± 0.12 (n = 15)	0.075 ± 0.016 (n = 6)

Table 9. *Activities of glycolytic and associated enzymes in H. vulgaris muscles*

Enzyme activities are expressed as μmol of NADH oxidized/min per mg of protein at 25°C and given as means $\pm$ S.E.M. of determination on n separate animals. Results taken from Trausch (1976*b*).

Enzyme	Claw muscle	Tail muscle
Phosphoenolpyruvate carboxykinase	0.002 ($n = 6$)	0.002 ($n = 6$)
Pyruvate kinase	0.88 ± 0.27 ($n = 15$)	1.46 ± 0.29 ($n = 18$)
L-Malate dehydrogenase (EC 1.1.1.37)	0.495 ± 0.044 ($n = 6$)	0.0656 ± 0.0083 ($n = 6$)
L-Lactate dehydrogenase	0.27 ± 0.12 ($n = 15$)	1.59 ± 0.48 ($n = 18$)

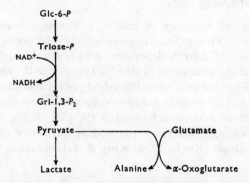

Fig. 3. *Lactate and alanine as end products of anaerobic glycolysis in Crustacea*

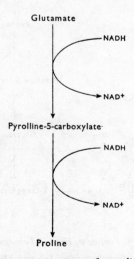

Fig. 4. *Glutamate as precursor for proline synthesis*

Table 10. *Proline oxidase activity from various tissues of E. sinensis*

Activities are given as variation of absorbance in 1 min/g wet weight of tissue. 2,6-Dichloro-phenol-indophenol is used as acceptor. Results from Vincent-Marique & Gilles (1970).

			Gills	
Hepatopancreas	Muscle	Nerve	Anterior	Posterior
0.112	—	0.037	0.062	0.362
—	—	—	0.075	0.312
0.012	—	0.037	0.025	0.275
—	—	—	0.037	0.237
—	—	—	0.100	0.375
0.175	—	—	0.087	0.287

produced at the triose phosphate step are utilized for the reduction of pyruvate or for the reductive amination of 2-oxoglutarate.

Proline as Chemical Energy Store

Glutamic acid is the precursor of proline following a sequence of reactions summarized in Fig. 4. Proline is an osmotic effector the concentration of which increases by a factor of 2 to 5, depending on the species, when the animal is submitted to a hyperosmotic stress (Table 3). This is also the amino acid released in larger amounts from the cells when the salinity of the medium decreases. When comparing the ability of various tissues of *E. sinensis* to oxidize proline, it is found that the greatest activity is located in the gills (Table 10) and more precisely in those three posterior pairs involved in the active absorption of salts from the dilute medium (Koch, 1954; King & Schoffeniels, 1969). This result

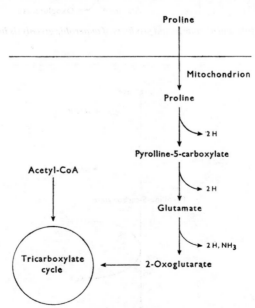

Fig. 5. *Proline as chemical energy store*
Complete oxidation of proline provides an additional 42 mol of ATP.

strongly suggests that, as found in some insects (Bursell, 1963, 1966), proline is a readily available source of energy. The oxidation of proline proceeds via the pathway indicated in Fig. 5. It is obvious that to be complete the oxidation of 2-oxoglutarate entering the tricarboxylate cycle must be accompanied by a stoicheiometric supply of acetyl-CoA. In this scheme, the complete oxidation of proline should provide an additional 42 mol of ATP. Together with its property of energy storage, proline could also be considered as an end product of anaerobic glycolysis since its synthesis from glutamate requires 4 reducing equivalents.

Pivotal Role of Glutamate Synthesis

The fundamental importance of glutamate dehydrogenase (EC 1.4.1.2; EC 1.4.1.3) stems from the fact that it is the most important if not the only way Crustacea have to introduce inorganic nitrogen as NH_3 into organic molecules (Schoffeniels, 1968a). By considering the standard oxidoreduction potential of the systems involved it is apparent that the enzyme favours the reductive amination rather than the reverse reaction. The oxidoreduction potential values for the systems $NADH/NAD^+$ and glutamate/2-oxoglutarate are respectively -0.32 V and -0.108 V indicating that the presence of any concentration of NH_3 or 2-oxoglutarate above the infinitesimal would strongly oppose the oxidative deamination of glutamate. As the synthesis of glutamate from its corresponding oxo acid is dependent on the availability of reducing equivalents as well as of that of NH_3, one may thus consider that a competition must exist between the various metabolic sequences needing reducing equivalents. As a matter of consequence one should find some kind of interaction between the respiratory chain and the synthesis of glutamate.

The results of O_2 consumption are already in favour of such a view since a low O_2 uptake parallels an increased synthesis of amino acids. More precisely many authors have pointed out the pivotal role of glutamate synthesis in the control of the energy metabolism of mitochondria (Ernster & Navazio, 1957; Tager & Slater, 1963; Slater & Tager, 1963; Papa et al., 1969a,b; Schoffeniels, 1968b).

As far as the euryhaline Crustacea are concerned, the results of Gilles & Schoffeniels (1965, 1968b) show that after the addition of NH_4Cl to the medium bathing an isolated nerve chain of crayfish one observes an increase in the intracellular pool of free amino acids and a decrease in the O_2 consumption. This is at variance with the results obtained with the lobster nerve chain in which the O_2 consumption is unaffected. Such a difference between the euryhaline crayfish and the stenohaline lobster indicates that a biochemical aspect of euryhalinity must be searched for in the control of the fate of reducing equivalents. Glutamate dehydrogenase is certainly involved in the process as demonstrated by the results of Tables 11 and 12. Addition of 2-oxoglutarate and NH_4Cl to the incubating medium of isolated axons of E. sinensis produces an increase in the free amino acid concentrations and a decrease in the O_2 uptake, a result confirming the previous findings of Gilles & Schoffeniels (1965) on the euryhaline crayfish. Thus an increase in the substrate concentration forces the glutamate synthesis, the reducing equivalent being diverted from the respiratory chain as demon-

Table 11. *Effect of 2-oxoglutarate and NH₄Cl on the amino acid pool of the isolated axon of E. sinensis*

Results are expressed as μmol/g fresh weight. For the experimental condition the incubation medium contained 10 mM-2-oxoglutarate and 1 mM-NH₄Cl. –, Not measurable; tr., traces. Two separate sets of measurements are given (Gilles, 1974a).

Amino acids	Amino acid content			
	Control	Experiment	Control	Experiment
Alanine	15.11	19.01	9.76	16.52
Arginine	6.70	–	4.90	10.26
Aspartate	43.64	65.29	30.73	73.47
Glutamate	7.09	10.42	5.49	12.00
Glycine	3.49	4.48	2.99	4.65
Histidine	tr.	tr.	tr.	0.34
Isoleucine	tr.	tr.	tr.	tr.
Leucine	tr.	0.40	tr.	0.28
Lysine	0.89	1.11	tr.	0.82
Methionine	tr.	–	–	tr.
Ornithine	tr.	0.38	tr.	tr.
Phenylalanine	–	tr.	tr.	tr.
Proline	6.33	10.37	5.03	10.01
Serine	1.49	1.91	1.20	1.74
Threonine	tr.	0.95	0.56	1.59
Tyrosine	–	–	tr.	tr.
Valine	tr.	tr.	tr.	0.29

Table 12. *Effect of 10 mM-2-oxoglutarate and 1 mM-NH₄Cl on the O₂ consumption of the isolated axon of E. sinensis*

The results were taken from Gilles (1974a).

Substrate	QO₂ (μl/h per mg dry wt.)	
—	3.64	2.87
2-Oxoglutarate (10 mM)	3.90	—
NH₄Cl (1 mM)	—	2.15
2-Oxoglutarate+NH₄Cl	2.16	1.38

strated by the lowering of the O_2 consumption. As a result of an increase in glutamate concentration, transamination reactions also increase, as well as the proline synthesis. Thus alanine, aspartate, glycine, proline, serine and threonine concentrations are higher.

Such events are only possible if key reactions in the processes involved are tightly controlled thus providing an integrated network of reactions leading to the nicely balanced mechanisms at play in euryhaline species. This aspect of the problem will be discussed later in the text.

Serine Hydro-lyase: Key Role in Deamination

Although serine is not on a quantitative basis an important osmotic effector, its key position in controlling the catabolism of amino acids is readily recognized

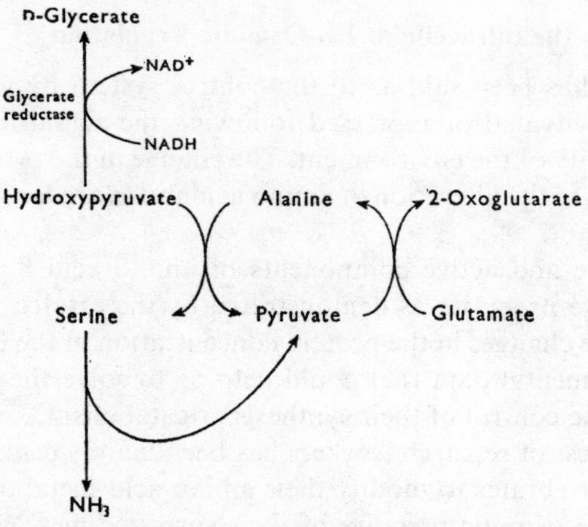

Fig. 6. *Key role of serine hydro-lyase in the deamination of amino acids in Crustacea*

Notice the importance of glyoxylate reductase in the control of hydroxypyruvate concentration.

by considering the scheme of Fig. 6. Serine is obtained by transamination reactions with hydroxypyruvate. It is catabolized to pyruvate and NH_3 through the activity of serine hydro-lyase (EC 4.2.1.13). On the other hand the availability of hydroxypyruvate is linked to the activity of glyoxylate reductase that catalyses the reduction of hydroxypyruvate to D-glycerate. Transamination of 2-oxo acids with glutamate as NH_3-group donor represents one of the main pathways in the biosynthesis of amino acids acting as intracellular osmotic effectors such as serine, glycine, alanine and aspartate. Any increase in glutamate concentrations should thus lead to an increase in the intracellular concentration of the above amino acids. However, owing to the presence of the enzyme serine hydro-lyase, the NH_3 group introduced at the level of the reductive amination of 2-oxo-glutarate is liberated into the medium. This situation is observed during a hypo-osmotic stress, when the intracellular concentration of the amino acids decreases (Table 13).

In relation to the suggestion that the properties of glutamate dehydrogenase favour reductive amination rather than the reverse reaction, it is certainly reasonable to propose that, together with glutaminase, serine hydro-lyase is responsible for the disposal of the amino group as NH_3.

Table 13. *NH_3 production in isolated axons of E. sinensis during osmotic stress*

Results are expressed in μmol/100 mg wet weight of axons or as μmol of NH_3 appearing in the total volume of medium per 100 mg wet weight of axons. The results are taken from Gilles & Schoffeniels (1969a).

	Hypo-osmotic stress		Hyperosmotic stress	
	Control	Experiment	Control	Experiment
Axon	1.841	0.404	2.978	0.159
Incubating medium	1.611	3.202	2.228	1.549

Biochemical Key to the Intracellular Iso-Osmotic Regulation

So far nothing has been said as to the control system by which the various mechanisms are activated or repressed following the signal introduced by the change in the salinity of the environment. The change in the properties of the cell membrane leading to the alteration of amino acid fluxes are far from being elucidated.

Both the passive and active components of amino acid fluxes are modified during the adaptive processes as demonstrated by the results of Gerard (1975). With respect to the changes in the protein concentration in the blood, we are still lacking the experimental data that could help us to solve the problem of their origin as well as the control of their synthesis or catabolism.

So far, the interest of research workers has been mainly centred on the ability of euryhaline invertebrates to modify their amino acid metabolism with respect to variations in the osmotic pressure of their surroundings. We shall therefore try to investigate the part played by different pathways of amino acid synthesis in the establishment of the amino acid pattern during the adaptation process and seek the primary cause for the modification of the activity of those pathways during adaptation to media having different salinities (Schoffeniels, 1971).

If in the incubating medium of isolated nerves from E. sinensis an increase in osmotic pressure is achieved by addition of sucrose to the dilute medium, nerves become electrically inactive within a few hours and the intracellular amino nitrogen concentration considerably decreases instead of undergoing an increase, as is the case when the rise in osmotic pressure is obtained by addition of sea water (Schoffeniels, 1960). This experiment clearly demonstrates that it is not the osmotic pressure *per se* that is responsible for the regulation of the amino acid concentration during the adaptation to concentrated media.

It is also evident from the experiments performed on isolated nerves (Schoffeniels, 1960; Gilles & Schoffeniels, 1969a) that part of the iso-osmotic regulation process is not under hormonal control since the isolated tissue is able to regulate, at least within certain limits, its amino acid concentration with respect to the osmotic pressure of the incubating medium. It can therefore be suggested that it is the modification of the ionic concentration to which the cell is submitted during adaptation which controls the mechanism responsible for the modification of the amino acid concentrations (Schoffeniels, 1960). In connexion with this hypothesis, Gilles & Schoffeniels (1964a, 1968b) have demonstrated that electrical stimulation or some substances, the action of which are classically explained by a modification in the intracellular ionic composition, have an effect on the synthesis of amino acids. To explain such a type of control two major possibilities exist. Either the ionic composition of the incubating medium acts through an intermediary substance on the amino acid metabolism or some ionic species control directly the activity of the enzymes involved in the amino acid metabolism. In the first interpretation, one would deal with the kind of mechanism generally used to explain a hormonal action, thus postulating the production of an intermediary effector such as cyclic AMP. As shown by the results of Table 14 there is a variation in the cyclic AMP content of various tissues when the animal is submitted to an osmotic stress. As to the presence of cyclic AMP in the tissues

Table 14. *Evaluation of 3′:5′-cyclic AMP content of various tissues of E. sinensis during osmotic stress*

The results were taken from Schoffeniels (1976).

Cyclic AMP content (pmol/g wet wt.)

	Fresh water (several months)	Fresh water (3 days)	Sea water (several months)	Sea water (3 days)
Muscle	270.9–397.5	—	609.8–734.3	—
	543–686	—	1417	—
	658.5–701.5	435–440	652.9–638	1102
Gills	475.3–492.9	—	1025.5	—
	631–626	—	564–577	—
	576.9–647.9	733–901	981–995	643–781
Nerve	1021.8	—	3575.6	—
	548	—	948	—
	582–648	359–443	1914–2089	864–1054
Hepatopancreas	225.8–326.4	—	311.8	—
	968	—	1227	—
	879–972.6	422–463	1953	798–810

of Crustacea, our results are in agreement with those of Bauchau *et al.* (1968). It is, however, the first time that variations in cellular content of cyclic AMP have been related to the cell volume regulation. Under the influence of a hyperosmotic stress the concentration of cyclic AMP increases, whereas the reverse is true when the animal is transferred to fresh water. According to Bauchau *et al.* (1968), cyclic AMP affects the phosphorylase activity according to the concentration used: *in vitro* a concentration of 2.5 μmol/ml inhibits the enzyme, whereas at lower concentrations, the enzyme is activated. If the results of Bauchau *et al.* (1968) can be extrapolated to our situation *in vivo* one should expect a stimulation of the enzyme activity under the influence of a hyperosmotic stress. This obviously makes sense since more keto acids are needed to meet the requirements of the volume regulation in a concentrated medium. However, we have to look for the signal that triggers the adenylate cyclase activity. On the basis of the results discussed so far we have no indication that the cyclase could be directly influenced by the change in the intracellular ionic composition. That part of the regulatory mechanism is maybe under hormonal control (see Scheer *et al.*, 1951). This would explain why isolated axons (*C. sapidus* or *E. sinensis*) submitted to a hyperosmotic stress have a decrease in volume amounting to 25%. There is no volume regulation within the next few hours unless cyclic AMP (1 mM) or dibutyryl cyclic AMP (1 μM) or plasma from an animal in the process of adaptation is added to the medium (J. F. Gerard, unpublished work). Even in the absence of volume regulation the amino acid pattern evolves and the increase in concentration observed cannot solely be accounted for by the shrinkage of the cell. This could perhaps indicate that besides its activating effect on glycogen hydrolysis, cyclic AMP could influence other mechanisms of cell-volume regulation still undiscovered or the turnover rates of glycine and proline.

As to the interpretation that the control of amino acid metabolism is mediated by inorganic ions, the results accumulated since it was first proposed (Schoffeniels,

1960) indicate a coherent picture of the situation. After it was clearly demonstrated that glutamate dehydrogenase is a key enzyme, because its catalytic properties are highly dependent on the ionic composition of the incubating medium (Schoffeniels, 1964b, 1965, 1966a,b, 1968a,b) together my collaborators and I have looked for enzymes involved directly or indirectly in the nitrogen metabolism, the activity of which could be affected by the ionic composition of the incubation medium. (Schoffeniels, 1970a,b,c; Schoffeniels & Bollette-Dugaillay, 1970; Bollette-Dugaillay & Schoffeniels, 1969, 1970).

Our studies have been performed on the following enzyme systems: (1) those involved directly in the amino acid metabolism such as glutamate dehydrogenase (EC 1.4.1.2), aspartate aminotransferase (EC 2.6.1.1), alanine aminotransferase (EC 2.6.1.2), serine hydro-lyase (EC 4.2.1.13), aspartate decarboxylase (EC 4.1.1.12) or glyoxylate reductase (EC 1.1.1.26); (2) those involved in the supply of the oxo precursors required for amino acid synthesis, such as succinate dehydrogenase (EC 1.3.99.1), isocitrate dehydrogenase (EC 1.1.1.42), malate dehydrogenase (EC 1.1.1.37 and EC 1.1.1.40), malate hydro-lyase (EC 4.2.1.2) and oxaloacetate decarboxylase (EC 4.1.1.3); (3) those involved in the fate of reducing equivalents in the intracellular medium such as lactate dehydrogenase (EC 1.1.1.27) or 3-glycerophosphate dehydrogenase (EC 1.1.1.8).

These experiments have been performed on tissue extracts from stenohaline and euryhaline species. They have been summarized in monographs (Schoffeniels, 1967; Florkin & Schoffeniels, 1965, 1969) as well as in two review articles (Schoffeniels, 1973, 1976). I shall therefore only cite the most salient facts that are relevant to my interpretation of the control of the amino acid pool in the cells of euryhaline Crustacea. Table 15 summarizes the most important data in relation to the effect of inorganic ions on the catalytic properties of enzymes located at crucial nodes in the metabolic network.

From Table 15 it can be seen that ionic composition of the medium has no effect on the catalytic activity of aspartate aminotransferase, at least within the limits of the so-called physiological range. It is thus reasonable to conclude that the rate of transamination is primarily controlled by the concentrations of the amino group donors as well as by that of the 2-oxo acid acceptors. In this respect the pivotal role of glutamate becomes evident since it is through the reductive amination of 2-oxoglutarate that NH_3 is introduced into amino acids. Since the activity of glutamate dehydrogenase is narrowly controlled by inorganic ions in such a way that an increase in catalytic activity up to 300% is observed at a salt concentration close to that found in the intracellular fluid of a marine Crustacea, I consider that the primary event in the regulatory process is situated at this level (Schoffeniels, 1964b, 1965, 1966a, 1968b). An increase in glutamate dehydrogenase activity must be accompanied by a larger supply of reducing equivalents. It is thus of great interest to observe an inhibition of lactate dehydrogenase and 3-glycerophosphate dehydrogenase as the salt concentration of the incubation medium increases (Schoffeniels, 1968b). Thus more reducing equivalents are made available for glutamate synthesis and accordingly O_2 consumption should decrease, as is observed experimentally.

The concentration of the NH_3-group acceptors must also increase. This situation seems also to be controlled by inorganic ions. First the rate of glycolysis

Table 15. *Effect of NaCl on the activity of various enzymes extracted from muscle of euryhaline Crustacea*

Results are given as percentage of the control activity taken as 100. The activity is given by (1) ΔE_{340} due to NADH oxidation for glutamate dehydrogenase, 3-glycerophosphate dehydrogenase, lactate dehydrogenase, malate dehydrogenase (EC 1.1.1.37) and glyoxylate reductase, ΔE_{340} due to the disappearance of oxaloacetate for aspartate aminotransferase and ΔE_{340} due to the decrease in fumarate concentration for malate hydro-lyase; (2) the quantity of pyruvate appearing after a 20min incubation period for serine hydro-lyase; (3) the amount of CO_2 produced after a 30min incubation period for aspartate or oxaloacetate decarboxylase. See also Schoffeniels & Gilles (1963), Schoffeniels (1964b, 1965, 1966a, 1968a,b, 1973, 1976) and Florkin & Schoffeniels (1969).

Enzyme	Species	NaCl concentration (mM) Enzyme activity							
		0	50	100	200	300	400	500	600
Aspartate aminotransferase (EC 2.6.1.1)	*A. fluviatilis*	100	100	100	95	—	102	—	—
Lactate dehydrogenase (EC 1.1.1.27)	*Maja squinado*	100	80	68	38	25	—	15	—
	E. sinensis	100	100	83	58	25	25	25	—
	A. fluviatilis	100	100	87.5	87.5	42	25	25	—
	Carcinus maenas	100	105	75	35	50	37.5	—	—
	Portunus puber	100	125	120	62	35	10	—	—
3-Glycerophosphate dehydrogenase (EC 1.1.1.8)	*E. sinensis*	100	—	300	280	240	160	120	100
	Carcinus maenas	100	—	305	283	250	200	133	100
	Portunus puber	100	—	212	225	206	169	131	94
	A. fluviatilis	100	—	366	315	217	195	183	116
Glyoxylate reductase (EC 1.1.1.26)	*E. sinensis*	100	—	91	14	9	6	3	2
	Carcinus maenas	100	—	30	16	8	6	4	3
	A. fluviatilis	100	—	30	15	7	6	—	3
Serine hydro-lyase (EC 4.2.1.13)	*A. fluviatilis*	100	54	32	14	—	—	—	—
	A. fluviatilis	100	63	38	3	—	—	—	—
Glutamate dehydrogenase (EC 1.4.1.2)	*H. vulgaris*	100	130	160	180	—	250	—	160
	A. fluviatilis	100	150	230	270	—	310	—	220
Aspartate decarboxylase (EC 4.1.1.11)	*H. vulgaris*	100	—	—	99	—	104	—	—
	A. fluviatilis	100	—	—	95	—	86	—	—
Malate dehydrogenase (EC 1.1.1.37)	*A. fluviatilis*	100	103	112	83	—	—	—	—
Malate hydro-lyase (EC 4.2.1.2)	*A. fluviatilis*	100	74	57	30	—	—	—	—
Isocitrate dehydrogenase (EC 1.1.1.42)	*A. fluviatilis*	100	87	73	63	—	50	—	29
Malate dehydrogenase (EC 1.1.1.40)	*A. fluviatilis*	100	68	58	51	—	43	—	31
Oxaloacetate decarboxylase (EC 4.1.1.3)	*A. fluviatilis*	100	96	86	72	—	65	—	58

goes up because of a greater availability of glucose thanks to the rise in cyclic AMP concentration. Also anaplerotic pathways for the synthesis of 2-oxoglutarate and oxaloacetate have been described in Crustacea. 2-Oxoglutarate is produced from intermediates of the hexose monophosphate shunt via arabinose (Gilles & Schoffeniels, 1969b), and oxaloacetate is obtained by carboxylation of pyruvate via the activity of both malate dehydrogenases. The first enzyme is inhibited by inorganic ions in the direction of pyruvate production from malate, thus indicating that malate production is favoured under the same conditions. The production of malate from oxaloacetate is also inhibited by inorganic ions suggesting a positive effect on oxaloacetate production via the activity of the second malate dehydrogenase. Moreover glyoxylate reductase, which catalyses the reduction of glyoxylate to glycollate or that of hydroxypyruvate to D-glycerate, is greatly inhibited by inorganic ions. This represents not only a gain in reducing equivalents but also provides more hydroxypyruvate for serine synthesis. Finally if the results obtained on *Mytilus californianus* (Gilles *et al.*, 1971) are applicable to the succinate

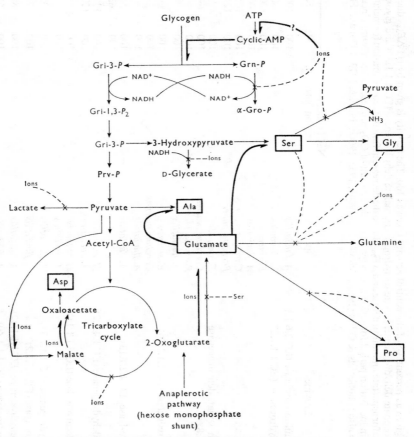

Fig. 7. *An integrated scheme of the effects of inorganic ions on the amino acid metabolism in Crustacea*

Note the central role played by glutamate dehydrogenase as well as by the enzymes controlling the use of reducing equivalents.

dehydrogenase of euryhaline Crustacea, one should postulate an inhibition of the tricarboxylate cycle at this level under the influence of inorganic ions thus leaving intact the upper part of the cycle functioning between malate and 2-oxoglutarate. The entries would then be malate (carboxylation of pyruvate), 2-oxoglutarate (from isocitrate or pentose cycle via arabinose) and acetyl-CoA.

The situation found in a euryhaline Crustacea when submitted to a hyperosmotic stress is summarized in the scheme of Fig. 7.

The initial stimulus is a change in the salt concentration of the environment. It is rapidly associated with a change in the blood and finally in the cell ionic content. Cyclic AMP synthesis is stimulated, thus providing the extra supply of the necessary carbon skeleton, and glutamate dehydrogenase is activated. More reducing equivalents are available because of the relative inhibition of 3-glycero-phosphate dehydrogenase, lactate dehydrogenase and glyoxylate dehydrogenase. The other enzymes are affected by the change in ionic composition as described above. As a result, O_2 consumption decreases and more alanine, proline, glycine and serine are formed enabling the cell to cope with the increased ion concentration in the blood. It should be emphasized that proline controls its synthetic pathway by feedback and that glutamine synthetase is inhibited by serine and glycine. Moreover glutaminase is activated and glutamate dehydrogenase inhibited by multivalent anions such as phosphate. We thus deal with a network nicely balanced thanks to the properties of key enzymes, the activities of which are controlled by various effectors among which the inorganic ions are amongst the most important.

Fate of Intracellular Amino Acids during a Hypo-Osmotic Stress

On the basis of the results presented in the preceding sections, one important question must be answered about the relative importance of the various ways of depleting the intracellular pool of amino acids. From observations of intra-cellular NH_3 production during hypo-osmotic stress, it may be concluded that deamination of amino acids does occur through the activity of various trans-aminases and serine hydro-lyase. Little is known, however, about the metabolic fate of the oxo acids produced. It is generally assumed that they are oxidized, thus explaining the increase in O_2 consumption and CO_2 production during hypo-osmotic stress (Chaplin et al., 1966, 1970; Huggins & Munday, 1968; Gilles, 1973, 1974a). This is at variance with my interpretation (Schoffeniels, 1968b) that the oxygen consumption is controlled by the availability of reducing equiva-lents. In other words the reducing equivalents produced at the level of the main energy substrate oxidation are accepted either by O_2 in a medium of low salinity or by oxo acids in a medium of high salinity.

There are indeed discrepancies between the experimental findings and the inter-pretation that oxo acids formed from amino acids are the main energy substrates.

It is certainly true that when isolated axons of C. sapidus are incubated in the presence of ^{14}C-labelled amino acids such as glutamate, aspartate, arginine, leucine, alanine or serine, under conditions of hypo-osmotic stress, one obtains an increase of 1.2–1.6 in $^{14}CO_2$ production depending on the amino acid used. However, if one relates the amount of amino acid oxidized to the total CO_2

output, it is apparent that the oxidation of the amino acid represents a maximum of 1% of the total CO_2 production. This is true whether one considers axons submitted to hyper- or hypo-osmotic stress. It is even more instructive to calculate the respiratory quotient (RQ) in the same experimental situations. Whatever the substrate used, or the osmotic stress, the RQ is always close to 0.75 (Schoffeniels, 1974, 1976). These results strongly indicate that oxo acids derived from the deamination of the osmotic effectors do not contribute much to the activity of the oxidative pathway followed by the main substrate that is responsible for the low RQ observed. These results obtained on isolated axons have been reproduced by Huggins *et al.* (1975) on whole animals such as *Carcinus maenas*.

An RQ as low as 0.75 could be taken as indicating that lipids are used as the main energy source, irrespective of the exogenous labelled substrate added. Though a determination of RQ cannot indicate precisely whether the substrate oxidized is derived from a limited group of molecules or is produced by a wide range of molecular species, it remains true, however, that a low RQ is indicative of oxidizable substrates with relatively low O_2 content, such as fatty acids. With palmitic acid (C_{16}) as sole substrate the RQ would be 0.696 whereas with hexanoic acid (C_6) it would be 0.75. On the basis of the results presented by Huggins *et al.* (1975), palmitic acid cannot be considered as the main substrate in the energy metabolism of the Crustacea. It thus remains to determine whether short-chain fatty acids such as butyric acid, valeric acid or hexanoic acid are preferential energy substrates.

What is then the fate of the oxo acids originating from the deamination of the osmotic effectors? An obvious possibility is that they enter the gluconeogenic pathway and are stored as osmotically inactive glycogen. This may well be the case in view of the effect of an osmotic stress on the intracellular cyclic AMP content.

It would then be reasonable to assume that in a medium of high salinity the oxidation of short-chain fatty acids proceeds through the Lynen cycle, but that a proportion of the reducing equivalents and acetyl-CoA produced is geared to the synthesis of oxo acids and amino acids. More oxo acids and reducing power would also be produced from glycogen hydrolysis according to the scheme of Fig. 7. This would lead to a decrease in O_2 consumption and CO_2 output because less short-chain fatty acids would be fully oxidized to CO_2 and water. By contrast, in a dilute medium, more short-chain fatty acids are fully oxidized so accounting for the increased O_2 consumption and CO_2 production. There would be no change in RQ, however, since the oxo acids derived from the deamination of osmotic effectors would serve for glycogen synthesis.

There is still another problem to be solved and it is that of the origin of the amino acids needed for the extra synthesis of blood proteins in a dilute medium. According to the results of Table 3 when *E. sinensis* is submitted to a hypo-osmotic stress there is a loss of $28.16 - 14.54 = 13.62 \mu mol$ of amino acids per 100 mg fresh weight of muscle. If one excludes from this loss that due to proline leaking out of the cell and being oxidized in the gills, we have $13.62 - 2.73 = 10.89 \mu mol$ of amino acids per 100 mg wet weight of muscle. Assuming that a specimen weighing 130 g has 36 g of tissue and an extracellular space of 20% of the total animal weight (Binns, 1969), i.e. around 24 g, the total amount of amino acids lost by the cells is

$10.89\,\mu\text{mol} \times 360 \simeq 4000\,\mu\text{mol}$. If the gain in the blood proteins is $30\,\text{g/litre}$ (see Table 6) a total of $720\,\text{mg}$ of protein would appear in the $24\,\text{g}$ of circulating fluid. Considering that the average molecular weight of an amino acid is roughly 100 a loss of $4000\,\mu\text{mol}$ of intracellular amino acids could account for $400\,\text{mg}$ out of the $720\,\text{mg}$ of protein appearing in the blood. Moreover, it can be estimated from the results of Florkin *et al.* (1964) that a Chinese crab of about $130\,\text{g}$ loses $0.8\,\text{mmol}$ of N_2 (as NH_4^+) into the medium when submitted to a hypo-osmotic stress. If all the NH_4^+ produced were related to the oxidation of proline lost by the cell this would amount up to $1.6\,\text{mmol}$ of proline. This value has to be compared with an intracellular loss of proline of about $1\,\text{mmol}$.

However, one should also compare NH_3 excretion with sodium uptake at the gills arising out of the adaptation to low-salinity medium.

The balance proposed above is far from being perfect, but it should be kept in mind that the calculations are performed on results obtained with different animals over the last 10 years. They, however, point to the need for a more refined study of the energy balance in relation to the amino acid metabolism as influenced by the salinity of the environment.

This work is supported by grant no. 790 from the Fonds de la Recherche Fondamentale Collective.

References

Bauchau, A. G., Merigeot, J. C. & Olivier, M. A. (1968) *Gen. Comp. Endocrinol.* **11**, 132–138
Binns, R. (1969) *J. Physiol.* **51**, 11–16
Bollette-Dugaillay, S. & Schoffeniels, E. (1969) *Arch. Int. Physiol. Biochim.* **77**, 493–500
Bollette-Dugaillay, S. & Schoffeniels, E. (1970) *Arch. Int. Physiol. Biochim.* **78**, 23–27
Boyland, E. (1928) *Biochem. J.* **22**, 362–380
Bricteux-Grégoire, S., Duchâteau-Bosson, Gh., Jeuniaux, Ch. & Florkin, M. (1962) *Arch. Int. Physiol. Biochim.* **70**, 273–286
Bursell, E. (1963) *J. Insect Physiol.* **9**, 439–452
Bursell, E. (1966) *Comp. Biochem. Physiol.* **19**, 809–818
Camien, M. N., Sarlet, H., Duchâteau, Gh. & Florkin, M. (1951) *J. Biol. Chem.* **193**, 881–885
Chaplin, A. E., Huggins, A. K. & Munday, K. A. (1966) *Biochem. J.* **99**, 42P–43P
Chaplin, A. E., Huggins, A. K. & Munday, K. A. (1970) *Int. J. Biochem.* **1**, 385–400
Cheng, S. C. & Mela, P. (1966) *J. Neurochem.* **13**, 281–287
Cole, K. S. (1932) *J. Cell. Comp. Physiol.* **1**, 1–9
Dendinger, J. E. & Schatzlein, F. C. (1973) *Comp. Biochem. Physiol. B* **46**, 699–708
Deyrup, I. (1953) *J. Gen. Physiol.* **36**, 739–749
Duchâteau, Gh. & Florkin, M. (1955) *Arch. Int. Physiol. Biochim.* **63**, 249–251
Duchâteau, Gh. & Florkin, M. (1956) *J. Physiol. (Paris)* **48**, 520
Duchâteau, Gh. & Florkin, M. (1961) *Comp. Biochem. Physiol.* **3**, 245–249
Duchâteau, Gh., Florkin, M. & Jeuniaux, Ch. (1959) *Arch. Int. Physiol. Biochim.* **67**, 489–500
Ernster, L. & Navazio, F. (1957) *Biochem. Biophys. Acta* **26**, 408–415
Florkin, M. (1956) in *Vergleichen Biochemischen Fragen: 6 Colloquium des Gesellschaft füi Physiologische Chemie*, pp. 62–99, Springer-Verlag, Berlin
Florkin, M. & Schoffeniels, E. (1965) *Stud. Comp. Biochem. Pap. Comp. Biochem. Meet. 1963* **23**, 6–40
Florkin, M. & Schoffeniels, E. (1969) *Adapted Molecules: A Molecular Approach to Ecology*, Academic Press, New York
Florkin, M., Duchâteau-Bosson, Gh., Jeuniaux, Ch. & Schoffeniels, E. (1964) *Arch. Int. Physiol. Biochim.* **72**, 892–906
Gerard, J. F. (1975) *Comp. Biochem. Physiol. A* **51**, 225–229
Gilles, R. (1973) *Neth. J. Sea Res.* **7**, 280–289
Gilles, R. (1974*a*) *Métabolisme des Acides Amines et Contrôle du Volume Cellulaire*, Thèse Agérgation, Université de Liège
Gilles, R. (1974*b*) *Life Sci.* **15**, 1363–1369

Gilles, R. & Schoffeniels, E. (1964a) *Biochim. Biophys. Acta* **82**, 518–524
Gilles, R. & Schoffeniels, E. (1964b) *Biochim. Biophys. Acta* **82**, 525–537
Gilles, R. & Schoffeniels, E. (1965) *Arch. Int. Physiol. Biochim.* **73**, 144–145
Gilles, R. & Schoffeniels, E. (1968a) *Arch. Int. Physiol. Biochim.* **76**, 441–451
Gilles, R. & Schoffeniels, E. (1968b) *Arch. Int. Physiol. Biochim.* **76**, 452–464
Gilles, R. & Schoffeniels, E. (1969a) *Comp. Biochem. Physiol.* **28**, 417–423
Gilles, R. & Schoffeniels, E. (1969b) *Comp. Biochem. Physiol.* **28**, 1145–1152
Gilles, R. & Schoffeniels, E. (1969c) *Comp. Biochem. Physiol.* **31**, 927–939
Gilles, R., Hogue, P. & Kearney, E. B. (1971) *Life Sci.* **10**, 1421–1427
Huggins, A. K. (1966) *Comp. Biochem. Physiol.* **18**, 283–290
Huggins, A. K. & Munday, K. A. (1968) *Adv. Comp. Physiol. Biochem.* **3**, 271–378
Huggins, A. K., Amrit, D. & Haworth, C. (1975) *Biochem. Soc. Trans.* **3**, 669–671
Jeuniaux, Ch., Bricteux-Grégoire, S. & Florkin, M. (1961) *Cah. Biol. Mar.* **2**, 373–379
King, E. N. (1965) *Comp. Biochem. Physiol.* **15**, 93–102
King, E. N. & Schoffeniels, E. (1969) *Arch. Int. Physiol. Biochim.* **77**, 105–111
Koch, H. J. (1954) in *Recent developments in cell physiology* (Kitching, J. A., ed.), pp. 15–27, Academic Press, New York
Leaf, A. (1956) *Biochem. J.* **62**, 241–248
Mudge, G. H. (1951) *Am. J. Physiol.* **165**, 113–127
Needham, A. E. (1957) *Physiol. Comp. Oecol.* **4**, 209–239
Opie, E. L. (1949) *J. Exp. Med.* **89**, 185–208
Papa, S., Tager, J. M., Francavilla, A. & Quagliariello, E. (1969a) *Biochim. Biophys. Acta* **172**, 20–29
Papa, S., Tager, J. M., Guerrieri, F. & Quagliariello, E. (1969b) *Biochim. Biophys. Acta* **172**, 184–186
Rand, R. P. & Burton, A. C. (1964) *Biophys. J.* **4**, 491–495
Robinson, J. R. (1960) *Physiol. Rev.* **40**, 112–149
Rothstein, A. (1959) *Bacteriol. Rev.* **23**, 175–201
Scheer, B. T. & Scheer, M. A. R. (1951) *Physiol. Comp. Oecol.* **2**, 198–209
Scheer, B. T., Schwabe, C. W. & Scheer, M. A. R. (1951) *Physiol. Comp. Oecol.* **2**, 327–338
Schoffeniels, E. (1960) *Arch. Int. Physiol. Biochim.* **68**, 696–698
Schoffeniels, E. (1964a) in *Comparative Biochemistry* (Florkin, M. & Mason, H. S., eds.), vol. 7, pp. 137–202, Academic Press, New York and London
Schoffeniels, E. (1964b) *Life Sci.* **3**, 845–850
Schoffeniels, E. (1965) *Arch. Int. Physiol. Biochim.* **73**, 73–80
Schoffeniels, E. (1966a) *Arch. Int. Physiol. Biochim.* **74**, 333–335
Schoffeniels, E. (1966b) *Arch. Int. Physiol. Biochim.* **74**, 665–676
Schoffeniels, E. (1967) *Cellular Aspects of Membrane Permeability*, Pergamon Press, Oxford
Schoffeniels, E. (1968a) in *Structure and Function of Nervous System* (Bourne, G. H., ed.), vol. 2, Academic Press, New York
Schoffeniels, E. (1968b) *Arch. Int. Physiol. Biochim.* **76**, 319–343
Schoffeniels, E. (1970a) *Arch. Int. Physiol. Biochim.* **78**, 135–139
Schoffeniels, E. (1970b) *Arch. Int. Physiol. Biochim.* **78**, 161–163
Schoffeniels, E. (1970c) *Arch. Int. Physiol. Biochim.* **78**, 461–466
Schoffeniels, E. (1971) in *Biochemical Evolution and the Origin of Life* (Schoffeniels, E., ed.), vol. 2, pp. 314–335, North-Holland, Amsterdam
Schoffeniels, E. (1973) in *Comparative Physiology: Proc. Int. Congr. Comp. Physiol. Acquasparta 1972* (Bolis, L., Schmidt-Nielsen, K. & Maddrell, S. H. P., eds.), pp. 353–385, North-Holland, Amsterdam
Schoffeniels, E. (1974) *Soc. Exp. Biol. London Meet.*, December
Schoffeniels, E. (1976) *Perspect. Exp. Biol.* **1**, 107–124
Schoffeniels, E. & Bollette-Dugaillay, S. (1970) *Arch. Int. Physiol. Biochim.* **78**, 307–312
Schoffeniels, E. & Gilles, R. (1963) *Life Sci.* **2**, 834–839
Schoffeniels, E. & Gilles, R. (1970) *Chem. Zool.* **5**, 255–285
Slater, E. C. & Tager, J. M. (1963) *Biochim. Biophys. Acta* **77**, 276–300
Tager, J. M. & Slater, E. C. (1963) *Biochim. Biophys. Acta* **77**, 227–245
Teal, J. M. & Carey, F. G. (1967) *Physiol. Zool.* **40**, 83–91
Tosteson, D. C. (1964) in *The Cellular Functions of Membrane Transport* (Hoffman, J. F., ed.), pp. 3–22, Prentice-Hall, Englewood-Cliffs
Trausch, G. (1976a) *Biochem. Syst. Ecol.* **4**, in the press
Trausch, G. (1976b) *Biochem. Syst. Ecol.* **4**, in the press
Vincent-Marique, C. & Gilles, R. (1970) *Life Sci.* **9**, 509–512
Whittam, R. (1956) *J. Physiol.* (*London*) **131**, 542–554

Biochem. Soc. Symp. (1976) **41**, 205–223
Printed in Great Britain

Adaptations of Enzymes for Regulation of Catalytic Function

By DANIEL E. ATKINSON

*Biochemistry Division, Department of Chemistry, University of California,
Los Angeles, CA 90024, U.S.A.*

Synopsis

1. Enzymes must not only be extremely effective catalysts, but must also be the operating components of very sensitive and sophisticated regulatory systems. Appropriate evolutionary adjustment of the properties of different enzymes causes the sites that bind typical metabolic intermediates to be only partially saturated *in vivo*, thus allowing flexibility for control by variation in ligand concentration. In contrast, sites that bind such coupling agents as the pyridine and adenine nucleotides seem to be virtually saturated. Thus the ratios, rather than the absolute concentrations, of these compounds are important in metabolic regulation. This type of response is essential to the function of these compounds simultaneously as thermodynamic energy transducers and modifiers in the kinetic regulatory system. 2. Reaction orders of two to four are frequently encountered. They appear to be essential to biochemical homoeostasis, which is the maintenance of nearly constant substrate concentrations at the expense of wide variation in flux rates. 3. The strategies of enzyme adaptation are general, but the actual adaptations of enzymes are highly specific, reflecting the place of the enzyme in a metabolic sequence, the place of the sequence in the metabolism of the cell and, in complex organisms, the function of the cell, and of the organ or tissue of which it is a part, in the organism. Patterns of adaptation must be almost infinitely varied. A few are presently known in outline, but probably none as yet in detail. Several examples are discussed.

Biological adaptation to environmental change occurs at all levels, from the behavioural to the molecular. At whatever level, adaptation involves two components. First, there are the slow genetic modifications by which a species changes its pattern of fitness, becoming simultaneously less well adapted to the old environment and better adapted to a new one. Such adaptations may occur either when conditions in the organism's habitat are changing or when members of the population are invading a new habitat in which conditions differ from those of the ancestral niche. Second are changes within an inherited repertoire of responses that allow individual organisms to adapt without genetic alteration to variations in their surroundings. All organisms can cope with at least some changes in the environment, and some species possess spectacular abilities in this respect. Adaptations of both types are relevant to this symposium, and both have been discussed. I will follow the lead of the organizers and consider both kinds of adaptations, essentially interchangeably. There is another kind of adaptation that underlies both of these and is more fundamental; the design features that are required for life, even in the absence of environmental change. A complex network of chemical reactions, such as metabolism, is inherently unstable. Adaptations of this last type, those that are necessary for biological homoeostasis, necessarily supply part of the basis for adaptations of the second type, those that confer ability to survive environmental change. To cite a simple example, I suspect that many other teachers, like me, point out to undergraduate classes that the control of biosynthetic sequences by product negative feedback ensures that an organism will not waste energy in making a compound that it can obtain more cheaply from the surroundings. Of course that is one consequence of feedback control, but a relatively trivial one. Even if amino acids were never encountered in the environment, an organism must necessarily regulate the production of each amino acid in such a way as to maintain reasonably constant concentrations. Life would evidently be impossible in the absence of effective stabilization of concentrations, and the most common immediate cause of cellular death is probably the excursion of one or more metabolite concentrations beyond the range within which the regulatory mechanisms of the cell can restore normal levels.

Having these essential feedback controls to regulate the concentrations of the end products of biosynthetic sequences, a cell can very easily gain the additional advantage of saving energy when end products are available. All that is necessary are transport systems with their controls adjusted to maintain concen trations of the products that are slightly higher than the concentrations at which the biosynthetic controls are set. The consequence would be nearly total inhibition of biosynthesis of any given product when it was available in the surroundings. This is not to propose, of course, that the total interaction between biosynthesis and uptake is as simple as this, but only to suggest a logically sufficient relationship that probably is a major component of the actual system *in vivo*.

Micro-organisms, with their enormous surface-to-mass ratios, are especially exposed to the environment. Very effective biochemical adaptations to environmental change must have evolved a billion years or more before any of the very interesting adaptations of more complex organisms that have been discussed in this symposium. Many of these adaptations must relate to properties of membranes, with which we are not concerned here. But life is chemical activity, and the primary stabilization of living systems and protection of their integrity from damage caused by the environment must be due to the evolved characteristics of enzymes. Even if the surroundings were constant, the existence of a living cell would depend on complex regulatory mechanisms. The non-constancy of the environment only exacerbates the problem.

The adaptations evolved by very early cells continue to provide for survival of contemporary micro-organisms. Many or most of them are also found in the cells of higher organisms, where they provide part of the biochemical regulatory mechanisms and supply the necessary basis for the physiological adaptations of higher organisms.

Even to be good catalysts, enzymes evolved properties quite unlike anything else in the world (and as far as we know, in the universe). A linear sequence of amino acids or other monomers would not be expected to form a compact globular structure. If such structures were formed, we would not expect their arrangement in space to be specifically determined by the sequence of monomeric units in the linear chain. Probably most random sequences of amino acids would not have this property, but enzymes are not random sequences. The first requirement of a polypeptide sequence that is to become an enzyme is that it must possess a combination of thermodynamic and kinetic properties that ensure its reliably and rapidly assuming the proper globular form. This globular protein must then have available a number of very specific responses to specific ligands. It must bind substrates at high and finely tuned affinities, must change conformation in such a way as to bring the reactants and participating functional groups on the enzyme into a reactive configuration (the transition state), and probably in most cases move parts of the enzyme aside to allow dissociation of the products. We are still far from any comprehensive understanding of the adaptations necessary to make an enzyme out of a protein, but in retrospect it seems surprising how long it took enzyme chemists to realize that changes in conformation must be involved in enzyme action, and that in adapting to become extremely effective catalysts proteins have evolved into machines with moving parts capable of specific changes of position that are essential to the catalytic action of the enzyme. Koshland's (1958) concept of induced fit, opposed initially by most workers in the field, may be considered to have ushered in the modern era of enzyme chemistry, as distinct from the period in which enzymes were usually thought of as rigid jigs. When a machine tool is forming a piece of metal, many or all parts of the machine are involved, not merely the parts that hold the work stock itself. Similarly, an enzyme reaction is catalysed by the enzyme molecule, not merely by the substrate-binding sites.

The conformational changes that accompany binding of the substrates and bring them into a reactive conformation may involve changes in positions of groups at many locations on the enzyme molecule, with accompanying changes in interaction energies among those groups, or between them and water molecules or surrounding ions. The 'free energy of activation' of the classical kinetic treatment will depend on the sum of all of these changes, some of which may involve groups far from the catalytic site. In relation to the size of an enzyme molecule and the number of interactions of many types in which its functional groups may participate, the free energy required to bring two or a few small reactant molecules into the reactive or transition state is negligible. It is thus not surprising that the apparent free energy of activation for enzyme-catalysed reactions can be extremely low. Presumably it can be decreased to whatever degree is functionally advantageous by evolutionary tailoring of concomitant conformational alterations for which the changes in free energy are negative. As one rather formalized example, the binding of the first substrate might be coupled to an otherwise unfavourable change in conformation, so that the negative free energy of binding overcomes the positive free energy of the conformational change. Binding of the second substrate could then allow the altered part of the molecule to relax back to its favoured conformation. If this relaxation was coupled to changes at the catalytic site that forced the substrates and participating enzymic functional groups into the reactive configuration, the likelihood of attainment of that configuration, and hence the velocity of reaction, might be enhanced to virtually any degree until diffusion or the rates of the conformational changes became rate-limiting. It is implicit in the induced-fit hypothesis of Koshland that the free energy of substrate binding is used to facilitate attainment of the reactive configuration. The extension suggested here is that part of the free energy of substrate binding may be stored in conformational alterations throughout the enzyme molecule. In effect, binding of one substrate (or of either substrate) may spring-load the enzyme in such a way that, when the molecule is triggered by the simultaneous presence at the catalytic site of all participants in the reaction, relaxation of the spring will force that site and the reactants into the transition-state configuration. One reason for the relatively large size of enzymes may thus be the need for a sufficiently extensive and specific three-dimensional structure, with many groups interacting with each other and with ions and molecules from the environment, to allow for delocalization of energy by storing it in generalized conformational changes that involve small changes in the free energies of many interactions. This proposed extension of the induced-fit hypothesis was suggested by reading a manuscript by Low & Somero (1975) dealing with interactions between enzymes and ions in the environment.

But merely being a catalyst, even of this very special kind, is not enough to allow for homoeostasis and the other characteristics of life. The additional requirements, even at the primitive level of our present understanding, seem uncomprehendibly restrictive. A complex cell must contain very many metabolites in solution. For this to be possible, all concentrations must be low. Since the concentrations of metabolites are dependent on the affinities of the enzymes that catalyse their reactions, these affinities must be high (the apparent Michaelis constants must be low) in order to hold the steady-state concentrations of metabolites at low values. Because metabolism is a highly branched network of reactions, with many compounds being acted on by two or more enzymes, high affinity for substrate is not enough; the affinities of different enzymes are interdependent, and it may be expected that a mutation that increased the affinity of an enzyme for its substrate would be as likely to be deleterious to the organism as a change in the opposite direction. Thus the pattern of substrate affinities of its enzymes must be one of the most fundamental of the adaptations necessary for a complex metabolism involving many reactions and many intermediates.

Biochemical homoeostasis depends on stabilization of concentrations by appropriate changes in fluxes through sequences, that is, in reaction rates. In most cases, the ordinary mass–action responses of metabolic reactions will be in the right direction to provide some stabilization. As the concentration of the metabolite rises, the extent of the increase will be limited by the consequent increase in the rates of the reaction or reactions for which it is a substrate. But for those enzymes whose action can be adequately described by the classical Michaelis treatment, the effective kinetic order is less than one at any substrate concentration, and it decreases with an increase in concentration. Thus such enzymes are not very effective in holding substrate con-

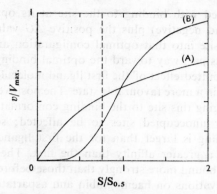

Fig. 1. *Relative velocity as a function of substrate concentration for a classical Michaelis enzyme (one with only one catalytic site or with no interaction between sites) (A) and a highly co-operative enzyme with four interacting catalytic sites (B)*

$V_{max.}$ indicates maximal velocity, and $S_{0.5}$ is the concentration of substrate at which the reaction velocity is half of the maximal rate. At the broken vertical line, the value of $S/S_{0.5}$ is 0.75.

centrations steady by increase in rate. This disadvantage has been overcome at a number of key points in metabolism by the evolution of enzymes that bind their substrates co-operatively, and hence catalyse reactions of kinetic order greater than one. As shown in Fig. 1, such enzymes respond much more sensitively to changes in concentration of their substrates, and thus more effectively limit swings in substrate concentrations, than do those that follow so-called first-order Michaelis kinetics. Enzymes that bind their substrates co-operatively are usually involved also in more complex regulatory interactions, but they contribute significantly to homoeostasis merely by their sensitive responses to variation in substrate concentration.

We should not underestimate the difficulty of the adaptations that lead to co-operative binding. Very specific long-range interactions are necessary. One of the simplest conceptual views attributes co-operativity to subunit–subunit interactions of very special types. According to this view, the monomers have an intrinsically high affinity for ligand. They also have a high affinity for each other. Their association is highly specific, and involves a distortion of the ligand-binding region in such a way as to decrease its affinity. That is, part of the negative free energy of association of the monomers is counteracted by the positive free energy change associated with forcing the conformation of the monomers away from the conformation that was previously most stable. When the first molecule of ligand binds, the value of its intrinsic negative free energy of binding is decreased by the need to change the binding site back to its high-affinity conformation, a change that now is associated with a positive free energy change because of the constraints between subunits. If a protein is to bind ligands co-operatively, the change in conformation associated with binding this first molecule of ligand must be transmitted to the other subunits in such a way as to increase their affinities for ligand. The free energy changes associated with these conformation alterations must also be positive. Thus the operational free energy of binding of the first molecule of ligand is the intrinsic ΔG associated with binding to the site at its optimal configuration (this will be large and negative) plus the positive ΔG values associated with forcing one binding site into that optimal configuration and pushing binding sites on other subunits part way toward the optimal binding conformation. As a result of this transmitted effect of the first ligand molecule, the second ligand molecule finds its site in a more favourable state. The positive free-energy change associated with changing this site to the binding conformation is thus smaller; also there are fewer unoccupied sites to be affected; so the net effective negative ΔG of binding is larger than for the first ligand. Thus the second molecule binds with a greater affinity than the first. The process continues, and each molecule is bound more strongly than those before. This hypothesis is consistent with observations on haemoglobin and aspartate transcarbamylase, the only two co-operatively binding proteins for which binding of ligand by monomers has been compared with binding by the native protein.

The first part of the proposal outlined above need not be generally true; it is not logically necessary that a pre-existing conformation favourable for binding be made less favourable by subunit association. The second part, on the other hand, must be true for all cases of co-operative binding, however much different enzymes may differ in the molecular mechanisms involved. Binding of the first molecules of ligand must facilitate binding of later molecules. Necessarily the net negative free energy of binding of early ligand molecules is decreased by the work done in forcing the binding sites that remain unoccupied into a conformation more favourable to binding. This is the essence of co-operative binding.

The concepts with which we deal are simple, but we must bear in mind that although it is easy to speak of changes in binding affinities caused by association of subunits or binding of ligand molecules, in the real world these changes are complex. They must be effected through highly specific changes in the positions of real atomic groupings in real three-dimensional space, and all of the mechanisms responsible for these sophisticated and specific changes must arise from spontaneous folding of a linear chain of amino acids.

In contrast with the co-operative binding that leads to high-order kinetics, some enzymes exhibit negative co-operativity with consequent kinetic orders less than that of the classical Michaelis enzyme. Koshland, especially, has discussed possible advantages of negative co-operativity. Here I want to consider an extreme case of decrease in kinetic order, a reaction with a negative order of reaction. ATP is a phosphoryl donor in the reaction catalysed by phosphofructo-kinase (eqn. 1).

$$\text{ATP} + \text{Fru-6-}P \rightleftharpoons \text{ADP} + \text{Fru-1,6-}P_2 \tag{1}$$

However, this reaction is a step in glycolysis and thus it participates in the production of ATP at the expense of oxidation of carbohydrates. It is therefore appropriate that it should be regulated, not as an ATP-utilizing reaction, but rather as a key step in an ATP-regenerating sequence. Accordingly, in the physiological range of ATP concentrations the reaction has effectively a negative kinetic order with regard to this reactant (Fig. 2). This is a remarkable feature of a reaction, which requires specific and sophisticated

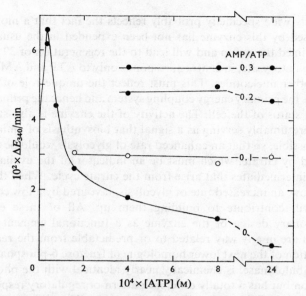

Fig. 2. *Inhibition of the reaction catalysed by yeast phosphofructokinase by ATP and counteraction of this effect by AMP*

The assay mixture contained 33 mM-Tris/HCl, pH 7.5, 3.3 mM-MgCl$_2$, 2.3 mM-glutathione, 0.1 mM-NADH, 0.28 mM-fructose 6-phosphate, excess of aldolase, triose phosphate isomerase and α-glycerophosphate dehydrogenase, ATP as indicated and enzyme. Where indicated, AMP was added to provide a molar ratio of AMP to ATP concentration of 0.1, 0.2 or 0.3, as shown on the appropriate curves. From Ramaiah *et al.* (1964).

adaptations of the catalyst. In this case, the affinity for ATP at the reactive site is very high, so that this site will always be essentially saturated under physiological conditions. One or more regulatory sites bind ATP, with lower affinity, and this binding causes a decrease in affinity for the other substrate fructose 6-phosphate at the catalytic site. This can be shown directly by the fact that ATP does not decrease the velocity of the reaction at high concentrations of fructose 6-phosphate, or by determining the $S_{0.5}$ values for fructose 6-phosphate (the concentration required for half-maximal velocity, corresponding to a Michaelis constant for a simple enzyme) at high and low concentrations of ATP. But as we will discuss shortly, the effects of variations in the concentration of a single adenine nucleotide are seldom, if ever, directly relevant to metabolic regulation. Enzymes respond to the ATP/ADP or ATP/AMP ratio, and responses to individual nucleotides seen *in vitro* have no direct bearing on anything of biological significance. In this case, the affinity for fructose 6-phosphate is modulated by the ATP/AMP ratio, as shown also by Fig. 2. When this ratio is held constant, the rate of the reaction is not changed significantly, even when the absolute concentration of ATP is increased by six to twelve times (Ramaiah *et al.*, 1964).

The functional adaptations of phosphofructokinase are extensive. Unlike most other kinases, this enzyme accepts GTP or ITP, and even the pyrimidine nucleoside triphosphates, as phosphoryl donors and can use them nearly as well as ATP.

8

This unusually wide specificity probably reflects the fact that a molecule of triphosphate used by this enzyme has not been expended in the usual sense, but rather has primed the pump and will lead to the regeneration of 39 or 37 mol of ATP. The regulatory site, in contrast, responds only to ATP and AMP, and is not affected by other nucleotides. This must reflect the unique role of the adenine nucleotides as the primary energy coupling system and hence the primary indicator of the energy status of the cell. The activity of the enzyme is also stimulated by NH_4^+ ions, presumably serving as a signal that biosynthesis of amino acids and proteins is possible, so that an enhanced rate of glycolysis would be appropriate. It is inhibited by citrate, which must be an indicator of the availability of the biosynthetic intermediates that arise from the citrate cycle. When these concentrations are low, an increased rate of glycolysis, favoured by a low concentration of citrate, will contribute to building them up. All of these effects result from evolutionary design of the enzyme as a functional element of the cell; none of them are in any way related to or predictable from the reaction itself. Phosphorylation of the next lower homologue of fructose 6-phosphate, catalysed by phosphoribulokinase, is chemically nearly identical with the phosphofructokinase reaction but has a totally different pattern of regulatory responses and in fact responds oppositely to the adenine nucleotides. The different responses of the two enzymes to adenylates is of course functionally significant, since phosphorylation of fructose leads to degradation of carbohydrate with regeneration of ATP, whereas phosphorylation of ribulose 5-phosphate leads to synthesis of carbohydrate with expenditure of ATP.

Probably the most general pattern of adaptation of enzyme behaviour in response to internal or environmental conditions of which we are aware are responses to modifier metabolites. These are ligands that, on binding at specific sites, change the affinity at the catalytic site for one or more substrates. The changes can be in either direction. As in the case of co-operative binding of substrate, part of the negative free energy of binding of the modifier molecule must serve to overcome the positive free energy of a specific conformational change that is coupled to modifier binding and alters the specific positioning of groups at the substrate-binding site. Again, the specificity for such interactions is extremely high.

Metabolic regulation typically involves the partitioning of a branchpoint metabolite between two or more enzymes that compete for it. Each enzyme, of course, directs the branchpoint compound into a metabolic sequence and commits it to a specific role. The outcome of the competition will depend on the relative amounts of the coupling enzymes and on their relative affinities for substrate. In the short range, the relative affinities are most subject to change, and their variations in response to variation in concentration of modifiers probably underlies most metabolic regulation.

Although we will not illustrate the point here, variation in affinity for substrate leads to much more sensitive regulation of partitioning between enzymes than could be attained by variation of maximal velocities by the same factor. Sensitive control is also facilitated by high-order reaction kinetics. Fig. 3 shows relative rates of two reactions of the same substrate, catalysed by competing enzymes, as a function of the concentration of the common substrate.

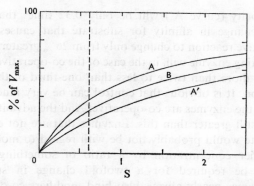

Fig. 3. *Relative rates of reactions catalysed by non-co-operative enzymes as a function of substrate concentration*

The $S_{0.5}$ values for the enzymes are: A, 1; B, $\sqrt{2}$; A', 2. The vertical broken line is at the concentration of substrate at which the velocity of the reaction catalysed by enzyme B is 30% of the maximal velocity.

The $S_{0.5}$ value for enzyme A is 1, and that for enzyme B is $\sqrt{2}$. Thus the two values differ by about 40%. The extent to which the rate of the reaction catalysed by enzyme A exceeds that of the reaction catalysed by enzyme B varies with substrate concentration, but never exceeds 40%. For example, when reaction B is proceeding at 30% of its maximal value (indicated by the vertical line), the rate of reaction A is 26% greater than that of reaction B. If, because of change in the concentration of a modifier, the affinity of enzyme A for substrate were decreased by a factor of 2, its response would be as shown by curve A'. Now reaction A would go at a velocity 78% of that of reaction B at the substrate concentration indicated by the vertical line. Fig. 4 shows the analogous case for a highly co-operative four-site enzyme. Again the $S_{0.5}$ values for enzyme A and enzyme B are 1 and $\sqrt{2}$ respectively. Now when the velocity of reaction B is 30% of its maximal value, reaction A will go 2.1 times as fast as reaction B. If $S_{0.5}$ for enzyme A is increased by a factor of 2,

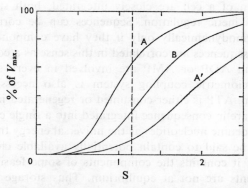

Fig. 4. *Relative rates of reaction catalysed by an enzyme containing four strongly interacting catalytic sites*

The details are the same as for Fig. 3.

the resulting velocity (curve A') will be only 0.32 times that of reaction B. Thus the same change in affinity for substrate that causes the velocity of the non-co-operative reaction to change only from 26 % greater to 22 % less than that of the competing enzyme will, in the case of the co-operative enzyme, change the velocity from more than twice to less than one-third of the velocity of the competing reaction. It is obvious that control can be very much more sensitive and precise when the enzymes are co-operative. And the advantage conferred by co-operativity is still greater than this. Enzymes that are not co-operative with regard to substrate would probably not be with regard to modifier either, so a change in modifier concentration by a ratio of something greater than 2 would probably be required for a twofold change in substrate affinity. Co-operative enzymes nearly always also bind modifiers co-operatively, and a twofold change in substrate affinity can be caused by a change of only a few percent in modifier concentration. Thus co-operative binding increases both the extent of change in substrate affinity that results from a given change in modifier concentration and also the extent of the change in substrate partitioning that results from a given affinity change. The multiplication of these two effects makes metabolite partitioning many times more sensitive than it could be in the absence of co-operative binding of substrates and modifiers. It seems probable that regulatory sensitivity of the degree necessary for systems as complex as contemporary cells would be entirely impossible in the absence of co-operative binding, and that such site–site interactions are therefore among the adaptations that were essential to the evolution of metabolism and life as we know it.

Interactions of the single-modifier type are appropriate in contributing to stabilization of the concentrations of such biosynthetic end products as amino acids, and in many other cases. They could not, however, provide for overall homoeostasis. It is not enough that individual metabolic sequences be regulated by the concentrations of their end products. Sequences must also be interrelated and correlated. A cell is far from a bag of enzymes, and equally far from being merely a number of sequences, each proceeding independently and regulated individually by its end product. The reaction sequences that make up the metabolism of a cell are closely integrated, both stoicheiometrically and in terms of kinetic regulation. Sequences can be correlated stoicheiometrically or thermodynamically only if they have components in common, and all metabolic sequences are correlated in this sense as a consequence of the fact that ATP and ADP or AMP are involved in every sequence. In this context, a stoicheiometric coupling system is also necessarily an energy-transducing system. ATP is either consumed or regenerated in every sequence, and all sequences are in consequence integrated into a single chemical energetic system, with the adenine nucleotides as the universal energy transducers.

A system may be said to contain potentially available chemical energy in only one sense, if it contains the components of some feasible reaction, and if those components are not at equilibrium. Thus storage of energy in the adenylate system depends on the concentrations of ATP, ADP, orthophosphate and water, and on the pH. At pH 7 and physiological phosphate concentration, the ATP/ADP concentration ratio is about 10^8 times the equilibrium ratio; thus the system contains a large amount of available chemical energy. The state of the system could be expressed by the ATP/ADP ratio, by the free energy of hydrolysis of ATP, or by the phosphoryl potential, which as usually defined is numerically identical with the free energy of hydrolysis. However, since the interactions of this system with metabolic sequences are stoicheiometric, a stoicheiometric parameter is more convenient for most metabolic purposes, just as the charge of a storage battery, which is a stoicheiometric parameter, is more convenient for most purposes than is the potential of a battery or the ratio among its components. The stoicheiometric parameter corresponding to a concentration ratio is a mole fraction. A mole fraction, being stoicheiometric, has the advantage of being linear with almost everything of chemical interest: the amount of energy stored in a system, the amount of reaction that has occurred or the amount of energy stored or delivered between any initial and any final state, and so on. A concentration ratio becomes infinitely large at the upper end of its range, and it is not linear with anything of chemical interest at any point within the range. The same disadvantage applies to the free energy of hydrolysis, or the phosphoryl potential, which is essentially the logarithm of the ATP/ADP concentration ratio plus a constant. This parameter has the added feature that, in addition to becoming infinitely large at one end of its range, it is infinitely large and negative at the other.

For two-component systems, like the primary electron-transfer system $NADPH/NADP^+$, a simple mole fraction is the appropriate stoicheiometric parameter for use in metabolism. Because there are three components in the adenylate system, all of which are involved in energy-transducing reactions, the parameter for use here must be slightly more complex. Through the action of adenylate kinase (eqn. 2), ADP can be converted into equal amounts of AMP and ATP.

$$2ADP \rightleftharpoons ATP + AMP \tag{2}$$

Thus two molecules of ADP are equivalent to one of ATP. Therefore the desired effective mole fraction of ATP, the parameter that linearly expresses the amount of metabolically available energy stored in the system, is the mole fraction of ATP plus half the mole fraction of ADP. By analogy with the corresponding parameter for a storage battery, this effective mole fraction is the charge of the adenylate system. To emphasize its metabolic significance, it has been termed the energy charge.

$$\text{Energy charge} = \frac{(ATP) + \frac{1}{2}(ADP)}{(ATP) + (ADP) + (AMP)} \tag{3}$$

The existence of a ubiquitous energy-coupling system greatly facilitates development of effective kinetic control and correlation between sequences. Indeed, it seems likely that this regulatory function was one of the main evolutionary factors in the development of a single central energy-transducing system. It is evident that if the various metabolic sequences are to be kept in step with each other the universal energy-transducing system must be very closely regulated. It is also obvious that sequences can be regulated in response to the cell's energy status only if there is some mechanism for sensing the energy status and regulating all metabolic sequences, whether energy-liberating or energy-consuming, accordingly. And, finally, it is obvious that the universal energy-transducing system must necessarily be an indicator of the energy status of the cell, and that a system that stabilizes the energy charge, slowing ATP-utilizing sequences and accelerating ATP-regenerating sequences when the charge is low and exerting opposite effects when the charge is high, will by the same token aid in maintaining functional correlation among all sequences. Thus once the function of the adenylates as the universal energy-transducing system is recognized, it is a nearly necessary conclusion that they should serve also as the primary basis of the kinetic correlation of metabolism. Any other arrangement would be far more complicated. It is very unlikely that a second system would be evolved to somehow sense the status of metabolic sequences and regulate them when the adenylates, in their stoicheiometric coupling role, already sense or, perhaps better, embody, the energy status of the cell.

Experiments *in vitro* with enzymes from glycolysis and the citrate cycle and with first enzymes in biosynthetic sequences have shown the expected responses. The results have been published over the past decade and need not be recited here. In summary, regulatory enzymes from catabolic sequences are maximally active at low values of energy charge, and their activities decrease as the charge nears the value of 1. Regulatory enzymes from biosynthetic sequences respond oppositely, showing low activity except at relatively high values of the charge. Both types of response curves have their midpoints at charge values of about 0.85. From these results, obtained *in vitro*, it appeared that the energy charge *in vivo* should be strongly stabilized in the vicinity of 0.85, since a decrease in the charge would be counteracted by the resulting increase in the rate of ATP-yielding sequences and decrease in the rate of ATP-consuming sequences and an increase in charge would lead to opposite effects. This prediction has been confirmed by many analyses for nucleotide concentrations *in vivo*, most of which were undertaken for other reasons than the determination of energy charge. It is now well established that the energy-charge values of normally metabolizing cells of most, if not all, types is between 0.75 and 0.95, and from the most recent and most careful experiments it appears that the range may be much narrower, perhaps between about 0.86 and 0.94.

From the ATP content and a measure of metabolic rate, for example the rate of oxygen uptake, the turnover time of the ATP pool may be estimated. It appears to be a second or less for various micro-organisms and a few seconds for some types of mammalian cells. Thus a small imbalance between the rates of utilization and regeneration of ATP, existing for a very short time, would lead to a large change in the value of the charge. The observed near-constancy of the charge under a variety of metabolic conditions testifies to the extreme sensitivity and speed of the regulatory system.

Regulation of the rates of metabolic sequences is a kinetic matter quite distinct from stoicheiometric energy coupling, which is thermodynamic. However, in some cases both functions may be served by interactions of ATP and ADP at the catalytic site of a kinase. The adaptations required for this double duty are relatively simple, as enzymic adaptations go. It must be remembered that energy storage in any chemical system is a function of the concentration ratios of its components. Thus it is the ratios among the adenine nucleotides, which can be expressed on the linear scale of energy charge, rather than the absolute concentration of any one of them, that should be sensed by regulatory enzymes. Experiments *in vitro* have fully confirmed this expectation. When the energy-charge response of an enzyme is determined and then the experiment is repeated with a larger or smaller adenylate pool (both pool levels being reasonably near the physiological concentration), the second curve has been found to superimpose on the first. That is, the value of energy charge is the controlling regulatory factor and the enzyme responds similarly to an energy charge of, for example, 0.8 whether the ATP concentration is 2 or 5 mM. We saw such a constant response to the ATP/AMP ratio over a wide range of absolute concentrations earlier for phosphofructokinase. Those experiments on phosphofructokinase were done before the birth of the energy-charge concept, and helped to lead to it. The same situation applies to the other general stoicheiometric coupling agents, the oxidation–reduction couples NADH/NAD+ and NADPH/NADP+. Again the ratio or mole fraction, rather than absolute concentrations, is the determining factor.

Fortunately, this desirable feature of sensitivity to ratios and insensitivity to absolute concentrations is relatively easily attained. All that is necessary is that the affinities for adenine nucleotide at the regulatory site, which, as noted above, may also be the catalytic site, should be high. Specifically, the $S_{0.5}$ values for ATP and ADP or AMP must be small in comparison with the physiological concentrations of these compounds. As a consequence, the adenylate-binding sites will be near saturation at all times, and binding can be affected only by changes in the concentration ratios of the nucleotides. Thus the type of response that is necessary for metabolic regulation can be obtained by evolution of nucleotide-binding sites with $S_{0.5}$ values lower than the physiological range of nucleotide concentrations, which in turn must be determined mainly by the $S_{0.5}$ values for feedback inhibitors in the purine-biosynthetic pathway.

Since the relative amount of each nucleotide bound will be proportional to the ratio $(S)/S_{0.5}$ for that nucleotide, it follows that the shape of the binding curve, and hence of the curve of kinetic response to changes in energy charge or mole fraction, will depend on the relative values of $S_{0.5}$ for ATP and ADP or for NAD+ and NADH, for example. Energy-charge response curves of the type actually observed for several kinases in biosynthetic sequences are generated directly by interactions at the catalytic site, merely by the $S_{0.5}$ value for ADP being smaller by a factor of 4 or 5 than the $S_{0.5}$ for ATP. Similarly, the regulatory responses seen for several dehydrogenases are generated by a higher affinity for the product nucleotide, NADH, than for the reactant, NAD+. Greater affinity for product than for reactant would not be expected if we considered the enzyme merely as a catalyst, since this situation necessarily leads to pronounced inhibition. It is precisely this product inhibition, however, that supplies the primary regulatory response for several enzymes of this type. The regulatory significance of simple $S_{0.5}$ values or dissociation constants has been discussed in more detail elsewhere (Atkinson *et al.*, 1975). Here the unexpected pattern need only be mentioned as an example of a strikingly effective adaptation that would make no sense in any terms except those of regulation and control.

It is easy to compare effects of absolute concentrations with those of mole fractions, ratios or energy charge *in vitro*, where both the pool level and the energy charge are totally under the investigator's control. Such comparisons are not so simple in the living cell, where the ATP concentration and the energy charge can be controlled only indirectly and imperfectly, and where they tend to vary in the same direction. We have been able to uncouple these variables partially by use of a mutant strain that cannot make adenine nucleotides because of a genetic block at the succinyl adenylate synthase step. When this strain is starved for adenine, the ATP concentration and the total adenylate pool fall by a factor of about 3 before there is any apparent change in energy charge. In such adenine-starved cells, the ability to incorporate radioactive leucine into protein was observed to parallel changes in energy charge much more closely than changes in the concentration of ATP (Swedes *et al.*, 1975). In perhaps the most general illustration of the relative sensitivities of intact cells to the energy charge and to the absolute concentration of ATP, growth was used as the criterion. In a chemostat, the adenine auxotroph was grown under adenine limitation (Fig. 5). At the right edge of the Figure, the culture

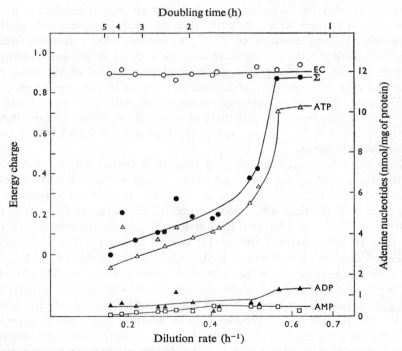

Fig. 5. *Adenine nucleotide concentrations and energy charge in adenine-limited cultures of an adenine-requiring strain of Escherichia coli (strain PC 0294)*

Growth rate in an adenine-limited chemostat culture was varied by the rate of medium addition. Each point represents the mean of three samples taken at hourly intervals after the cell density had reached a steady state. Curve identifications: EC, adenylate energy charge; Σ, adenylate pool (sum of ATP, ADP and AMP). From Swedes *et al.* (1975).

was growing without limitation at a rate equal to that of batch cultures in adequate medium. When adenine became limiting (moving to the left), the intracellular concentrations of ATP and of the other adenine nucleotides fell by over 50% with very little effect on growth rate. As the adenine limitation became progressively more stringent, the intracellular adenylate concentration fell further to about 30% of the normal concentration. Growth, although slow, continued. In contrast, the value of the energy charge did not vary by the minimum amount that we could have detected, 2 or 3%. This result confirms rather dramatically that in growing cells, as in experiments *in vitro*, the regulatory enzymes and the whole regulatory complex of the cell respond much more sensitively to energy charge than to the absolute concentration of ATP. Because of this sensitivity of response to the energy charge (and to its component ratios), the value of the charge is stabilized even in the face of drastic changes in absolute concentrations. Also because of this same sensitivity, growth seems not to occur at energy-charge values outside a very narrow normal range, although it is seen over a wide range of ATP concentrations. This response pattern of intact cells must depend entirely on the near-saturation of all important adenine-binding regulatory sites, including the dual-purpose catalytic and regulatory sites that have been studied *in vitro*. It is thus a striking conformation both of the expectations based on the chemical logic of energy transduction and of the predictions based on observations *in vitro* of sensitivity to energy charge and relative insensitivity to the absolute concentrations of the nucleotides.

My discussion so far has dealt with quite general types of enzyme adaptation. There must, of course, be many highly specific adaptations of individual enzymes. We have observed properties of adenylate deaminase from liver and ascites-tumour cells that we believe are part of one such highly individual regulatory system, and one with an interesting built-in self-limitation.

We have observed that a decline in energy charge is usually accompanied by a decrease in the total adenylate pool concentration. The same relationship is evident from the results of analyses for adenine nucleotides from cells or tissues of various types reported from several laboratories. The decrease in energy charge was caused in some cases by starvation and in others by putting a sudden heavy load on the adenylate system, for example by supplying a phosphoryl acceptor such as fructose or deoxyglucose. In some cases an accompanying increase in IMP, inosine or hypoxanthine suggested that the missing adenylates had been deaminated. These results led directly to the hypothesis that AMP deaminase is activated by a slight fall in the energy charge. Removal of AMP would tend to raise the value of the energy charge, since the concentration of AMP appears only in the denominator. It would also tend to raise the ATP/ADP ratio, since removal of AMP would unbalance the adenylate kinase reaction $(2ADP \rightleftharpoons ATP + AMP)$ and cause conversion of ADP into ATP to be favoured. This control would not be helpful in the long run because of the small amount of adenylates present in a cell, but it might well limit the depth of the negative spike of energy charge in such emergency situations, thus preventing the charge from falling to possibly lethal values, and 'buying time' during which fundamental regulatory mechanisms can adjust to meet the new situation. We delayed testing

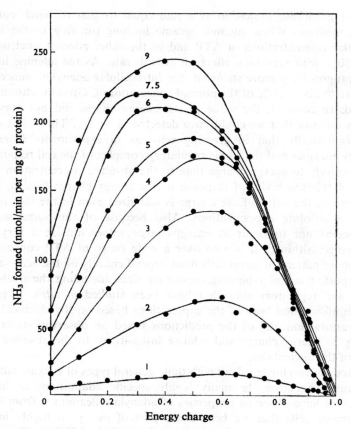

Fig. 6. *Response of AMP deaminase from ascites-tumour cells to variations in adenylate energy charge at different adenine nucleotide pool sizes*

Reaction mixtures contained 1–9mM total adenine nucleotides, as indicated by numbers identifying the curves at energy-charge values specified on the abscissa. NH₃ produced was measured by coupling with the glutamate dehydrogenase reaction. From Chapman *et al.* (1975).

this hypothesis for a year or two, however, because ATP had been reported to be a positive modifier for the deaminase. This report appeared to weaken our hypothesis, since we proposed that the enzyme should be activated when a sudden drain on ATP causes a fall in the energy charge. Stimulation by ATP seemed, at first thought, to be incompatible with that proposal. When we finally looked at the energy-charge response of the enzyme, however, a rationalization for the ATP effect was immediately obvious. The results also reinforced our belief that results of experiments in which only one of the adenine nucleotides is varied are likely to be of no value in terms of metabolic chemistry, and may often be seriously misleading. The response of AMP deaminase to variations in the value of the energy charge at various total adenylate pool concentrations is shown in Fig. 6. For the moment consider only one of the curves; let us say the top one. A regulatory site of the enzyme appears to be, under physiological conditions, nearly saturated with ATP. Thus the decrease in ATP concentration as the energy charge falls has initially little effect on the rate of deamination. The marked increase in AMP concentration, together with a higher-than-first-order response of reaction velocity to AMP concentration, is responsible for the sharp increase in rate of AMP removal as the charge falls. This removal will, as noted above, limit the extent of the decline in energy charge. The cell buys time at the expense of part of its adenylate pool. But if the responses at all pool concentrations were as shown by one of the upper curves, the cell would be in danger of losing its total adenine nucleotide pool in a very short time. This would doubtless be lethal, and a mechanism that protects against a momentary sharp drop in energy charge will be of little use if it kills the cell in the process. The significance of the ATP effect and of the specific degree of affinity for ATP that has evolved at the regulatory site is seen when the curves of Fig. 6 are compared. At a charge of 0.7, for example, the rate of deamination is nearly the same at pool values from 9 down to 4mM. As the pool concentration falls below the normal physiological range (which is probably about 4–6mM), the decrease in ATP concentration causes a sharp decline in the activity of the enzyme, so that the pool is protected against total depletion. The results presented in Fig. 6 were obtained with enzyme from ascites-tumour cells. Earlier experiments with the liver enzyme led to similar results (Chapman & Atkinson, 1973), which related very well to results obtained with intact livers after fructose perfusion or injection by Woods et al. (1970) and Raivio et al. (1969).

Escherichia coli and several other bacterial species seem not to contain AMP deaminase. These organisms, however, contain an enzyme, AMP nucleosidase, that splits AMP to adenine and ribose 5-phosphate, which in turn has not been observed in eukaryotic cells. The reactions catalysed by these two enzymes are chemically unrelated, and the products are different and clearly do not serve similar metabolic functions. The only thing the enzymes have in common is that they both remove AMP. They will therefore have similar effects on the adenylate energy charge and the size of the adenylate pool. It thus seems to me to be extremely interesting that the response of AMP nucleosidase to variation in energy charge is very similar to that of AMP deaminase, and that this enzyme, too, is strongly stimulated by ATP (Schramm & Leung, 1973). When rather unusual properties of an enzyme seem reasonably interpreted as having evolved to protect, by removal of AMP, against large and rapid falls in the energy charge, the case for the interpretation is strengthened by discovery of another enzyme from a different class of organisms that removes AMP by a different reaction, but exhibits the same unusual regulatory responses. Whether or not the explanation proves to be correct in detail, we think this type of mechanism may illustrate the types of specific adaptations of enzymes to environmental change that will be found in increasing numbers as biochemists become able and willing to look for them.

In conclusion, I would like to say a few words in praise of evolution. Enzymes are not just interesting objects for study, like stones or crystals. They are tightly designed mechanisms, and their meaning lies wholly in their evolved functional design. All metabolic chemists, of course, accept evolution, but rather few, it seems to me, think of metabolism first and foremost in terms of evolutionary teleology. Probably most enzyme kineticists have heard of evolution and approve. But many of them still deal with enzymes in terms of inappropriate classical rate equations, which submerge the actual properties of these delicately designed mechanisms in formal velocity constants that are quite unrelated to biological function.

I assert that metabolic chemists and enzyme kineticists unnecessarily work in self-imposed strait jackets when they fail to base their work solidly and explicitly on evolutionary concepts. In that assertion I am repeating the gist of a published lecture by Hans Krebs (1954), which profoundly affected my own scientific outlook a number of years ago. Research is to a large extent a process of blundering about in a dark room hoping for the best, and we need any help we can obtain. In any field of biology, the concept of evolutionary teleology is the best guide available. Probably even the most classical metabolic chemists would insist that they work within an evolutionary context. But a perfunctory *ad hoc* suggestion of functionality in discussing properties that were come upon in studying an enzyme or reaction for its own sake is far different from planning an experimental approach on the basis of a thorough-going acceptance of evolutionary teleology. I urge young workers to adopt the latter approach.

It is undeniable that the adenine nucleotides play an essential role in transduction of chemical energy. It is also undeniable that chemical energy is a matter of concentration ratios. Therefore, if we accept any degree of evolutionary teleology or biological functionality, it follows that concentration ratios and mole fractions of the adenylates must be of great importance. But this concept seems extraordinarily difficult for many metabolic chemists to accept. I have been told far more times than I wish to remember that the way to do valid experiments is to vary one concentration at a time. But if one wishes to obtain results that may be helpful in attempts to understand metabolic function and control *in vivo*, nothing could be less relevant than experiments in which effects of the components of coupling systems, such as the adenine and pyridine nucleotides, are tested individually.

In rejecting a manuscript that we submitted for publication a few months ago, the editor of a journal sent us the report of a referee who commented 'These authors obscure their data through the artifice of energy charge.' He wanted to know the concentrations of individual nucleotides, and especially asked about ADP. Would he, I wonder, respond similarly in his non-professional life? If a mechanic told him that the reason why his car failed to start was the low charge of the battery, would he answer 'Come now, don't obscure the situation. What is the content of lead dioxide?'? The functional complexity of a living cell is enormously greater than that of an automobile, let alone an automobile battery, and we should attempt in studying cells to use logic of at least as high an order as is commonplace in a repair shop. Yet this reviewer's comment is typical of many that we have received from other referees and members of seminar audiences. The concept that individual concentrations are real, whereas ratios or mole fractions are somehow artificial, dies hard. Because it is incompatible with a functional evolutionary view of energy metabolism and with the thermodynamics of energy transduction, it must die eventually.

A regulatory enzyme is a marvel of functionality and miniaturization, combining as it does within a single molecule a number of sensors or transducers, a non-linear processing system for handling the information from the transducers, and a controlled reactor. The signals to which most enzymes respond are intracellular, but many of them reflect, or are affected by, external conditions. Thus a cell, merely by existing and metabolizing, is constantly adapting to environmental conditions. Properties of enzymes that are essential to metabolic interrelation and correlation cannot be separated from those that are involved in adjustment to environmental change, since many features contribute importantly in both ways. Taken together, these types of responses are perhaps the most characteristically biological attributes of a living organism, and they underlie its existence as a programmed, persisting, and regulated entity distinct from the random and unregulated non-living environment.

References

Atkinson, D. E., Roach, P. J. & Swedes, J. S. (1975) *Adv. Enzyme Regul.* **13**, 393–411
Chapman, A. G. & Atkinson, D. E. (1973) *J. Biol. Chem.* **248**, 8309–8312
Chapman, A. G., Miller, A. L. & Atkinson, D. E. (1976) *Cancer Res.* in the press
Koshland, D. E., Jr. (1958) *Proc. Natl. Acad. Sci. U.S.A.* **44**, 98–104
Krebs, H. A. (1954) *Bull. Johns Hopkins Hosp.* **95**, 45–51
Low, P. S. & Somero, G. H. (1975) *Proc. Natl. Acad. Sci. U.S.A.* **72**, 3305–3309
Raivio, K. O., Kemomäki, M. P. & Mäenpää, P. H. (1969) *Biochem. Pharmacol.* **18**, 2615–2624
Ramaiah, A., Hathaway, J. A. & Atkinson, D. E. (1964) *J. Biol. Chem.* **239**, 3619–3622
Schramm, V. L. & Leung, H. (1973) *J. Biol. Chem.* **248**, 8313–8315
Swedes, J. S., Sedo, R. J. & Atkinson, D. E. (1975) *J. Biol. Chem.* **250**, 6930–6938
Woods, H. F., Eggleston, L. V. & Krebs, H. A. (1970) *Biochem. J.* **119**, 501–510

Author Index

Numbers in italics refer to pages in the references at the end of each chapter

225

Subject Index

A

Accidental hypothermia, 61, 63, 101

Acclimation, 1, 2, 38

Acetate, as invertebrate fermentation product, 17, 133, 136, 153, 165

Acetylcarnitine, 121

Acetylcholinesterase,
adaptation of binding site, 4–7
temperature and substrate binding by, 34–36

Activation enthalpy and entropy, 33, 40, 41–42

Activation free energy, of temperature-adapted enzymes, 33, 38–41

Actomyosin adenosine triphosphatase, insect, 129, 130

Adaptations,
acclimatization and, 1
anaerobiosis, to, 133–167, 169–178
diving animals, in, 11, 12–14, 169–178
enzymes, of, 3–31, 33–42, 205–223
insects, for flight in, 111–131
molluscs, for anerobiosis in, 133–167
muscle metabolism, of, 3–31, 76–80, 93–94, 111–131, 169–178
oxygen availability, to, 3, 9–31, 133–167, 169–174
salinity, with respect to, 167–204
substrate cycles and, 61–109
temperature, to, 24, 33–42, 43–60, 61–109, 175–176

Adenine nucleotides, *see also under specific names*, regulation of enzyme function by, 211–223

Adenosine 3′:5′-cyclic monophosphate (cyclic AMP), 80, 86, 93, 188, 196, 197, 201

Adenosine deaminases, adaptations of, 8

Adenosine diphosphate (ADP), mitochondrial regulator, 125, 126, 127

Adenosine monophosphate (AMP), enzyme regulation and, 115–117, 214–222

Adenosine monophosphate deaminase, *see* Adenylate deaminase

Adenosine monophosphate nucleosidase, 221

Adenosine triphosphatase,
insect flight-muscle, 129
myofibrilar, 10

Adenosine triphosphatase—*continued*
sodium ion translocation and, 87
temperature adaptation and, 46, 50–52, 56

Adenosine triphosphate (ATP),
anaerobic molluscs, in, 134–136, 158–159
enzyme regulator, as, 116–117, 214–222

Adenosine triphosphate–pyruvate phosphotransferase, *see* Pyruvate kinase.

Adenylate cyclase, 11

Adenylate deaminase, 219–221

Adenylate energy charge, *see* Energy charge.

Adenylate kinase, 86, 215

Adipose tissues, substrate cycles in, 74–78, 86–90

ADP, *see* Adenosine diphosphate.

Adrena trimmerana, 82

Adrenaline,
glycolytic activation by, 11
substrate cycles, effect on, 75–76, 79, 93, 94

Alanine,
anaerobiosis, role in, 18, 133, 136, 137, 146–150, 152–154, 155, 159, 160, 161, 163–165, 173, 174, 190–192
salinity, role in adaptation to, 181–192, 201

Alanine aminotransferase, 14, 20, 21, 24, 190, 198

Alcoholic hypothermia, 61, 63, 99–101

Alcoholic intoxication, 99, 100

Aldolase, diving-animal muscle, 12, 13

Alloxan-diabetes, 76

Amino acids, *see also under specific names*,
catabolism in anaerobic marine invertebrates, 17, 18
salinity involvement in adaptations, to, 179–204
substrate cycle with proteins, 98
temperature adaptation, transport and, 43, 44, 46–47, 52–58

AMP, *see* Adenosine monophosphate.

Amplification, in metabolic regulation, 61, 62, 67–72

Anaerobic–aerobic transition, in diving animals, 13–14

231